APPLYING S88

APPLYING S88

BATCH CONTROL FROM A USER'S PERSPECTIVE

JIM PARSHALL
AND LARRY LAMB

Notice

The information presented in this publication is for the general education of the reader. Because neither the authors nor the publisher have any control over the use of the information by the reader, both the authors and the publisher disclaim any and all liability of any kind arising out of such use. The reader is expected to exercise sound professional judgment in using any of the information presented in a particular application.

Additionally, neither the authors nor the publisher have investigated or considered the effect of any patents on the ability of the reader to use any of the information in a particular application. The reader is responsible for reviewing any possible patents that may affect any particular use of the information presented.

Any references to commercial products in the work are cited as examples only. Neither the authors nor the publisher endorse any referenced commercial product. Any trademarks or tradenames referenced belong to the respective owner of the mark or name. Neither the authors nor the publisher make any representation regarding the availability of any referenced commercial product at any time. The manufacturer's instructions on use of any commercial product must be followed at all times, even if in conflict with the information in this publication.

Printed July 2006

ISBN-10: 1-55617-703-8
ISBN-13: 978-1-55617-703-3

Printed in the United States of America.

ISA
67 Alexander Drive
P.O. Box 12277
Research Triangle Park
North Carolina 27709

Library of Congress Cataloging-in-Publication Data is available.

LC #99-42402

For our parents

Table of Contents

FOREWORD

To grossly paraphrase Nietzsche, serving on standards committees either kills you or makes you stronger. After six years of bimonthly international meetings involving hundreds of batch professionals, costing the sponsoring companies millions of dollars, accommodating thousands of end-user comments, boosting frequent flyer accounts to record highs, and leveling whole forests to produce the twelve intermediate drafts required to reach consensus, the ISA S88 Committee produced ANSI/ISA-S88.01-1995: Batch Control Systems, Part 1: Models and Terminology. The standard promised to unify the world of flexible process manufacturing with a common framework, finally allowing engineering, operations, vendors, and customers to use a common language to describe, design, and deliver agile plants.

While momentous, clear, and enlightening to the hard-core committee members, the standard proved in practice to be more of a *reference* than a *guide* to automation engineers. There were a few who vowed to write application guides to supplement the standard, but most approaches cover a superset of the S88.01 material, treating the many aspects of batch automation necessary to successfully architect and manage large teams. However, there is a need for strong examples and case studies that breathe life into the theory. That is where this book comes in.

While attending the Rockwell Consumer Products Roundtable a few years ago, I ran into Jim Parshall for the first time. He was holding court on the finer points of making ice cream, waving his arms, and genuinely having a good time explaining the process to those around him. I was instantly struck by his outgoing and enthusiastic attitude about applying computers to flexible manufacturing. He moved throughout the crowd, challenging whomever he encountered to a discussion on segmenting a plant, architecting a PLC, or laying out a graphic display. His combination of technical expertise and dynamic, engaging style appealed to the attendees and helped them understand his message. This book is written in the same style, telling the story of Ben & Jerry's first foray into the world of agile manufacturing. Jim and Larry breathe life into the technical details of S88.01, and the result is both entertaining and informative. With the standard as a companion, this book serves as a cookbook for success, guiding the S88 "newbie" through the perilous terrain of justifying, specifying, designing, and deploying a truly flexible process manufacturing system.

June 1999
Michael Saucier
Chairman and Founder of Sequencia Corporation

ACKNOWLEDGMENTS

Many authors acknowledge the help of others and often state how important everyone's contribution was to the effort. After this project, we now completely understand how important this section is. A whole lot of people advised us, lent a hand, or otherwise cheered us on during our batch control project and while we wrote this book. Our names may be on the cover, but there was a much larger team involved.

Most of the effort needed to write this was stolen from quality time with our families, and so it is only fitting that we first thank Georgianne and Evan Parshall (Evan was born just as we completed the first manuscript draft) and Deb, Erica, and Sarah Lamb. Their love, patience, and understanding gave us the time and ambition to accomplish this.

This book wouldn't have much value if we didn't upgrade the mix-making batch control system at the Ben & Jerry's St. Albans plant. The lead mix maker, Althea Sherwood, believed in us all the way and was a driving force in making the system better. Chad Larivee and Mark Kimball, the other two mix makers, patiently worked with us and showed the same faith as Althea. Ellyn Ladd, Production Manager, and Janet Norcross, Process Supervisor, gave us the time and support necessary to enhance the operations. Dan Carver, Controller, and Scott Beaudin, Cost Accountant, were key customers; helped us assemble a successful proposal; and worked with us diligently to ensure that the system delivered as promised. Gary Epright, Engineering Manager; Wendy Yoder, Plant Manager; Drake Wallis, Corporate Director of Manufacturing; and Bruce Bowman, Senior Director of Operations, let us use technology to improve plant operations. Sue Ketcham taught us volumes about process manufacturing. We returned the favor by helping her understand the world of modular automation (and linear algebra and logarithms). She's now supporting the batch control system at St. Albans (among her other many duties). You may have heard it before, but Ben Cohen and Jerry Greenfield are two real guys. We both knew them, and on one occasion Ben and Jerry bought beers for us at a local microbrewery. Of course, we must acknowledge our good friend and fellow engineer Roland Wilhelm. (He's a mechanical engineer, but we don't hold that against him—most of the time.) The three of us made quite a trio at that plant. Roland's unrelenting insistence on perfect project management, starting with exceptional requirements gathering, kept us on our toes throughout our tenure in St. Albans. He's truly one of a kind, and we miss working with him.

We had help from many people in industry as well. These next four people spent many hours reviewing our manuscript and suggested many great improvements. (So blame them for any mistakes ...) Michael Saucier, founder of Sequencia, guided us in the principles of S88, always honestly shared his views, and was kind enough to write the book's foreword. Tom Fisher, Technology Manager at

Lubrizol and considered the "father" of S88, kept us from straying too far from the intent of the standard. Karen Tipp, Industry Specialist at Rockwell Automation, is always a great source of encouragement and enthusiasm. Her experience implementing OpenBatch helped us clarify issues. Finally, Rick Mergen, Senior Engineer at Lubrizol and the first Chairman of the S88 committee, also provided great insight.

Lynn Craig of Manufacturing Automation Associates, and Chairman of the S88 committee when we wrote the book, is a wealth of information and phraseology. He read the manuscript on his own time and provided great feedback. Wayne Cantrell from Siemens helped us with PCS 7 information. Steve Ryan, Dan Hartnett, John Clark, and Chuck Fortner from Rockwell; Bart Winters and Don Clark from Honeywell; Bob Nelson from Siemens; Joe Hancock from Eutech; Bruce Sanchez from AspenTech; Mike Kolba from Foster-Wheeler (and the Chairman of the World Batch Forum while we wrote this); Roddy Martin from AMR; Asish Ghosh from ARC; Paul Nowicki from Sequencia; and Niels Haxthausen from Novo Nordisk Engineering all lent us a hand in one important way or another.

Lou Bendle from Sequencia showed us the light on OpenBatch and how we much we could benefit from it. John Parraga from Sequencia was the most-excellent consultant who introduced us to the world of phase logic. We also cannot forget the Sequencia product support team, as support can sometimes make or break a product. This group is first class. Thanks Barry, Tamara, Danny, Mark, and John.

In the middle of writing this book, we both ended up leaving Ben & Jerry's to pursue other ambitions. We want to thank Jim Newton and Bill Derochers of Oakes Electric for their support and encouragement. There are many people we want to thank at Eli Lilly. Paul McKenzie and Rae Marie Crisel ensured that no issues held us back and supported us all the way. Dave Adler went above the call of duty by reading the entire manuscript during a flight to Singapore. Jeff Owen and Jim Wiesler helped us better understand the world of distributed control systems.

Finally, we will not forget, nor can we emphasize enough, our appreciation to ISA for making this book possible. Kate Fern was the acquisitions editor who got us in the door and helped us refine the proposal. After Kate moved to get married, we had the fortune of working with Robert Rubin, a superb editor. Robert knew how to walk the fine line between ensuring that we told our story the best way and letting us keep our style and intent. He handled the role extremely well, and we sincerely appreciate all of his effort. Once the final manuscript was submitted, we had the pleasure of working with Shandra Botts, the production manager. Authors get really antsy about getting a book in print once the manuscript is finished. She did phenomenal work. We also want to thank Joice Blackson for all of her help in coordinating the proposal and manuscript reviews, as well as all of her efforts during the editing process.

Introduction

You probably remember reading a chapter from an engineering or math textbook and thinking to yourself, "I'm sure the author knows what he's talking about, but I'm just not getting it." So you read the chapter again. And you still didn't get it. You chugged or repeatedly sipped your favorite caffeinated beverage. Ah, but that didn't work either. So you highlighted the chapter with three or four different colors to outline what you thought were the important notes or equations. Your impending headache made you swallow a couple of aspirin with another jolt of caffeine. Maybe you just gave up and went to bed.

Well, as embarrassing as this may sound, reading the S88.01 standard reminded us a little bit of our college days. We know the SP88 members certainly have the qualifications to write such a spec and worked long and hard producing it. In fact, during the past several years we have had the pleasure of meeting several of them. Unfortunately, the nature of any standard or regulation prevents its authors from providing lengthy examples and details for interpreting it. Since standards serve countless companies in many different industries, they must be written to cover the broadest ground or they would never be published. (We defy you to tell us that *immediately* after reading any section of the FDA's "Good Manufacturing Practices" you understood *exactly* what the FDA meant.) But hey, let's not make any excuses: S88 just wasn't sinking in.

As we reviewed S88 and spoke with industry experts, including vendors, committee members, and other users, we took notes to strengthen our understanding of the standard. Our installation of RSBatch to control Ben & Jerry's mix-making operation in 1998 was a very strategic project for the company because RSBatch was a key component to improve manufacturing efficiencies and information handling. Integrating RSBatch with Ben & Jerry's existing process control system required a lot of planning, so we continued to take notes before and during the installation.

From our notes during that project we wrote this book. (We didn't dare write it from memory.) It's about what we think S88 means and how we used *our* interpretation to redesign our existing batching system. Our hope is that you'll only need one highlighter to capture the important points. We didn't have the good fortune to serve on SP88, so we don't consider our solutions to be definitive. In other words, we may not have implemented S88 exactly like others, but what we did worked well for us. In fact, we're proud to say that the system, including the human-machine interface (HMI), was designed, developed, and retrofitted into existing hardware and software in fewer than five hundred person-hours and that the very first batch of mix created with the new system was successfully used in finished product. (For any reader without an automation background, *HMI* is a manufacturing term that refers to a graphical user interface or "GUI." Some people may use the term *operator interface*. Folks who want to give away their age use the terms *operator interface terminal* or *OIT*.)

By now, you may think that we're using the terms *S88*, *S88.01*, and *S88* somewhat interchangeably. No, these aren't typos. (Hey, let's give our editor some credit, after all.) Each of these has a very distinct meaning: S88 is the overall standard name, S88.01 is part one of the standard (it has two parts), and SP88 is the name of the committee that authored the standard. These terms are explained in more detail early in Chapter 1.

How We Chose S88 and RSBatch

The Ben & Jerry's plant in St. Albans, Vermont, started running its mix making system in the summer of 1995. All batching functions were controlled via relay ladder logic (RLL) in Allen-Bradley PLC-5s and PanelViews. (Personal computers were not part of the process control system in 1995.) Once we eliminated the initial bugs the system was quite stable, as long as ingredients and processing steps stayed the same. That first batch control system quickly helped turn the St. Albans plant into the company's volume leader.

However, as a growing company committed to remaining competitive in the marketplace, Ben & Jerry's is constantly introducing new products and reformulating existing products. While many of these products used existing base mixes, some required new mixes. Well, you guessed it: many of these new mixes required new ingredients and new processing steps.

After introducing several new mixes, we concluded that our existing PLC-only mix-making control logic wasn't flexible enough to handle all of our new products. In addition, our method for collecting data was fairly primitive. Operators were tired of writing all recipe ingredients out by hand, along with target quantities, actual quantities, and the raw ingredient tanks used in each batch. (It was rumored that Ben & Jerry's was the largest purchaser of stainless steel clipboards in northern New England.) St. Albans had significantly reduced mix-making manufacturing variances since its start-up, but we knew we could do better. However, without more automatic and accurate data collection, even trying to perform a simple analysis of our process became a chore.

So, we set out to find a better way to run our mix-making system. We noticed something peculiar when reading batch control books: they didn't discuss using ladder logic to run and manage recipes. At first, we just figured the books were written from a perspective that favored distributed control systems (DCS), in which scripting-type languages are used for batch control.

However, while reading *Batch Control*, edited by A. E. Nisenfeld and H. Leegwater (ISA, 1996), we noticed the S88 standard being mentioned prominently throughout the text. We also couldn't help noticing the popularity of S88 in various trade magazines, so we decided to do a little research on S88.01 and ordered the standard from ISA. While we couldn't grasp all the concepts initially, we both thought following a standard for batch control was at least worth some investigation.

Meanwhile, while making a presentation at Allen-Bradley's Consumer Products Roundtable in March 1996, Jim met Michael Saucier, the founder of Sequencia (then known as PID). Sequencia created and is marketing OpenBatch, a PC-based batch control software package that follows the S88 standard. About that time, Rockwell Software announced that it had licensed OpenBatch and would be marketing it under the name "RSBatch." (Other companies, including Honeywell, Siemens, and AspenTech, have also since licensed OpenBatch and integrated it into their batch solutions.)

There are several "S88-aware" solutions in industry today, including products from Sequencia, Rockwell, Wonderware, Intellution, Siemens, Honeywell, Fisher-Rosemount, Moore Products, ABB, Yokogawa, and GSE. After reviewing the alternatives, we concluded that OpenBatch was the best solution for us.

Ben & Jerry's formed an alliance with Rockwell Automation in 1993 in which Rockwell would be Ben & Jerry's exclusive provider of critical control system components, such as PLCs, operator interfaces, discrete control panel components, frequency drives, and motor starters. Ben & Jerry's relationship with Rockwell Automation was very strong, and so Rockwell Software's decision to license OpenBatch strengthened our favorable opinion of the product. OpenBatch can work with all kinds of different control systems, and Allen-Bradley PLCs can work with many S88 batch packages. Because of our commitment to our alliance with Rockwell Automation and our satisfaction with that relationship, we decided to purchase RSBatch. (It's also easier to justify projects that include products and services from alliance partners.) However, to appeal to a wider audience (in hopes of higher royalty income), we've written this book to educate users who are working with any S88-aware solution. Furthermore, we'll also use the term *OpenBatch* instead of *RSBatch*.

Before we go any further, we do need to make one thing clear: we're not saying batches can't be controlled exclusively using ladder logic or that they shouldn't be (after all, the St. Albans plant used ladder logic for three years and has left the original PLC code in place as a backup). We're just saying it ended up not being the best way for us.

Besides retrofitting an existing control system with new software (versus installing an entirely new system), we did something else that was not common: we designed and implemented the system ourselves. Plenty of very qualified system integrators and consulting companies, including Rockwell and Sequencia, could have installed it for us. But we had the resources available and the personal interest to do it ourselves. What we're going to tell you—and what the examples we're going to use will show you—is not what we read, saw, or managed. They are what we *did*.

What You Should Read Depends on Who You Are

We believe this book can help educate just about anyone in an organization that uses batch control. In the following table we have listed suggested chapters to read based on your role at your company. We believe these apply if you work for a manufacturing company or if you're working at a vendor, an original equipment manufacturer (OEM), or a consulting firm.

If you are:	Think about reading:
An automation or controls engineer	The whole kit and caboodle
A project engineer or project manager	Almost the whole kit and caboodle: skip Chapter 12
An automation or controls technician	Chapters 1 through 8 and 11 through 14
An operator	Chapters 1 through 8 and 13 and 14
An information technology (IT) systems analyst in manufacturing	Chapters 1 through 10 and 13 and 14
An engineering or IT supervisor	Chapters 1 through 7 and 13 and 14
A mid-level manager and above (including a company executive)	The first two parts of Chapter 1 and all of Chapter 14

How We Chose to Organize This Book

So this is the story of how we implemented an S88-aware batch control system. In the first chapter, we begin discussing batch manufacturing so that we'll all start on the same page, literally and figuratively. Still in Chapter 1, we introduce S88 as a concept and a philosophy, throw entity-relationship (E-R) diagrams and sequential function charts at you, and describe our mix making process. Chapters 2 and 3 discuss critical project activities, like gathering customer requirements, selling the concept (getting money), managing the project, and scrounging dinners from vendors. Starting with Chapter 4, we dig into the meat (or tofu for you vegetarians out there) of S88.01. For those of you who have read the standard, please be patient with us. We do not follow its thought progression exactly. We'll introduce S88 models in Chapter 4, talk about recipes in Chapters 5 and 6, discuss equipment control in Chapters 7 and 8, and review important aspects of information handling in Chapter 9. So that you get your money's worth, we're also going to discuss specifying a batch control system in Chapter 10, designing phase logic in Chapter 11, and writing phase logic in Chapter 12. Chapter 13 is about starting your new system, including dealing with validation activities. We wrap things up in Chapter 14.

We didn't implement every detail of S88 in St. Albans because we didn't need to. But for this book we have also worked with industry experts to fill in the gaps. If you're not an OpenBatch (or RSBatch or Total Plant Batch) user, don't fear: you will learn and profit hugely from reading this book. We didn't write it to be a replacement user manual for OpenBatch, RSBatch, or Total Plant Batch. For those of you using batch solutions from Intellution or Fisher-Rosemount, this book will be more helpful to you than you may think. The core functionality of their products is very similar to OpenBatch.

You just can't please everybody, so we know there will probably be at least three groups of people who aren't going to like this book. The first group will be S88 committee members, who'll be disappointed because we're not going to interpret the standard exactly the way they intended. The second group includes engineers who implemented S88 differently than we did but who at least *think* they implemented their solution "the right way." The third group includes people much smarter than us, who'll find our interpretation of S88 offensively simple.

On the other hand, committee members who authored S88, engineers who have already implemented S88, and those who believe they truly understand the standard probably don't need this book. We hope you do—that you really have no idea what S88 is all about or at least you would like a guide to using it. After reading this, maybe you'll think we're both S88 geniuses and hire us as consultants for obscene fees.

On with the show.

1

Basic Concepts

No milling around in this book; we're jumping right in. So hang on tight and enjoy the ride.

Batch Manufacturing

Manufacturing operations can be generally classified into one of three different processes: discrete, continuous, and batch.

Discrete processes involve the production of things. A part or a specific quantity of parts in a group moves from one workstation to another, gaining value at each location as work is performed. In a discrete process, each thing or part maintains its unique identity. Often parts are combined to create products or other parts, but each new product or part maintains its unique identity. A great example of a discrete manufacturing process is the production of automobiles. Think conveyors, robots, nuts, bolts, and torque wrenches when considering discrete processes.

Continuous processes involve the continuous flow of material through various processing equipment. Once a continuous process is operating in a steady state, the goal is to produce a consistent product no matter how long the operation may run. The production of gasoline is often thought of as a prime example of a continuous process. Once started, refineries do not want to shut down. When looking at a continuous process, you'll see pumps, valves, instrumentation, and larger processing equipment (such as a cracking tower).

According to the S88 standard, a *batch process* is defined as follows: "a process that leads to the production of finite quantities of material by subjecting quantities of input materials to an ordered set of processing activities over a finite period of time using one or more pieces of equipment." So, instead of a continuous flow that can go on for days or weeks, batch processing involves limited quantities of material called—are you ready for this?—*batches*. By the nature of the process, batch manufacturing is *dis*continuous. That is, you start with some raw material, do something with it, send it on its way, and start all over again with some new raw material.

Batch manufacturing is also not discrete. There are no things that you can easily separate or identify. Sure, you can place a portion of a batch into some specific container, like a bottle of soy sauce, but that doesn't make the process discrete. If you combine a whole bunch of uniquely stamped gas caps in a box and mix them up, you can still identify each one individually. You can individually mark bottles of soy sauce, but the sauce *inside* the bottle is still part of the same batch and cannot be distinguished from one bottle to the next. The distinguishing factor of a product like soy sauce is the batch or lot from which it was bottled. That's why you'll see some type of batch or lot identifier printed on the cap or label.

We've dealt with batch processes throughout our lives. Our mothers made brownies in batches. We wash clothes in batches. (The clothing in the washing machine may be uniquely distinguishable, but washing is a batch process nonetheless.) The word *batch* is a noun, but it is also a verb. To "batch" is to apply a batch process.

Producing a product consistently is always a top priority. Once a process is repeatable, you can then work on issues like reducing cost, waste, or both. Consistently producing products with batch manufacturing is especially tricky. Unlike continuous processes, which may run for a long time, brand-new batches are created often. And unlike both discrete and continuous processes, it may not be possible to determine if the batch is being made correctly *while it is being made.* You may have to wait until the batch is complete before checking. If it's a bad batch, you can try to correct it (which costs time and money) or trash it (which *really* costs time and money). The point is to make a batch right the first time.

So, the control of the batch process (or as we automation engineers like to call it, *batch control*) is a very important aspect of batch manufacturing. Wow, what a lead-in to S88!

What Really Is S88?

First of all, for the purposes of our book, here's the golden rule regarding S88 and SP88, which also applies to all ISA standards and committees:

- *S88 is the standard* (*S* stands for "Standard").
- *SP88 is the committee that wrote it* (*SP* stands for "Standards & Practices").

Now that you know this, don't embarrass your friends, your family, or your company by using S88 or SP88 in the wrong context.

The S88 committee was formed to address batch control issues after companies grew sick and tired of four basic problems:

1. No universal model existed for batch control.
2. Users had a very difficult time communicating their batch processing requirements.
3. Engineers found it very hard to integrate solutions from different vendors.
4. Engineers and users had a difficult time configuring batch control solutions.

All of these problems led to expensive batch control systems that often did not meet all of the needs of the users and were difficult to maintain. So the committee put together a road map that outlined a two-part standard. ANSI/ISA-S88.01-1995 (the standard number) and *Batch Control Part 1: Models and Terminology* (the standard title) was released on October 23, 1995; S88.00.02, *Batch Control Part 2: Data Structures and Guidelines for Languages* had not yet been officially released by ISA or ANSI when we wrote this book. However, to gain valuable insight into S88.00.02, we activated the "mole" that we planted in the committee many years ago. Our source gave us important and accurate advance information about S88.00.02 that we have included in this book. Most of that information will be focused on data structures, which are discussed in Chapter 9. The portion of S88.00.02 that deals with language guidelines seems to be leaning more toward a vendor focus and is not quite as important to the user. We made a difficult decision not to include language guidelines in this book; however, we can always write another book.

We will often shorten our reference to the standard to "S88" (not S88.01 or S88.00.02). We've already had more than one committee member kindly inform us that the official standard names are ANSI/ISA-S88.01-1995 and S88.00.02. Okay, okay, we agree; but we're trying to save some ink here!

Ahem! Moving on ...

S88 isn't just a standard for software, equipment, or procedures; it's a way of thinking, a design philosophy. Understanding S88 will help you better design your processes and manufacture your products. Leveraging the knowledge and experience contained in the standard will enable you and your customers to identify your needs better, make recipe development easier, and help reduce the time it takes to reach full production levels with a new system or for each new product. By following the concepts explained in S88, you could improve the reliability of your operations and reduce the automation life-cycle cost of your batch processes, including lowering the initial cost of automating your operations. How does S88 do this?

- *S88 isolates equipment from recipes*—When the code to run equipment and the code that defines a product recipe are in the same device (such as a programmable logic controller [PLC] or a distributed control system), the two different sets of code eventually become indistinguishable and in some cases inseparable. Every additional ingredient and process improvement can require many person-hours to modify the software. Documenting such a system is also extremely tough. This makes recipes difficult, if not impossible, to maintain. (And let's not mention *altering* the process.) If recipes are kept at a higher level, as in S88-aware systems, they are more flexible. The person who knows what process changes are required—process engineers or lead operators—can make the changes directly. Control systems experts are not needed. (Whoa! So much for job security. But who wants to do "maintenance" work like this anyway?)

- *S88 provides guidelines on how to recover from abnormal events*—Recovering from abnormal events is one of the most difficult parts of batch control. In many non-S88 installations, automatic recovery is not implemented, and operators and/or engineers are needed to get equipment and the recipe back in sync. Since S88 uses a standard set of states, it provides engineers with an opportunity to consider how abnormal events should be handled when designing the system. Recovering from abnormal or unexpected events was one of our main considerations when we designed an S88 solution for Ben & Jerry's.

- *S88 helps you track historical data*—When there is a problem with a product and someone has to go back through plant data to try to determine the cause, it can be difficult to understand what was being done in the process at a given time. S88-aware software tracks the state of the batch in a log. Some companies integrate this process log into "historian" software packages.

- *S88 makes gathering requirements from customers easier*—S88's common terminology and models help you know what questions to ask. This will make it more likely that the system you install is the system your customers wanted. How a product is manufactured is often as important as the ingredients it is made of. S88 helps you better define the manufacturing process.

- *S88 makes conveying requirements to vendors easier*—Again, common terminology and models will help reduce the number of surprises you and your vendor spring on each other. Working with more than one vendor is also easier. If everyone follows S88, you are much more likely to have better success integrating products from different vendors. In the long term, as vendors become more comfortable with S88, you should experience quicker turnaround time on systems and projects.

- *S88 modularity makes possible better return on investment*—For those of you with large batching operations or several plants producing similar products, you can also benefit from the reusability of recipes and equipment phases that S88 makes possible. If your recipes are written properly, you can duplicate equipment functionality with minimal changes, which will significantly reduce the time you need to implement subsequent projects. With S88, recipes are also more transportable between sets of equipment or between plants.

- *S88 design concepts will make validation easier*—A good S88 design will allow you to validate procedures and equipment independently. This simplifies the commissioning process, which typically translates into a faster start-up and faster ramp-up to full production capability.

These seven points can translate into direct, tangible benefits. The following list shows possible improvements in terms of typical metrics that a batch manufacturing plant may capture. As you can see, some of these are directly related to others.

1. Reduced batch cycle time
2. Increased production run rate
3. Higher number of batches per year
4. Shorter changeover time
5. Lower cost to make a batch
6. Improved batch yield or consistency
7. Reduced raw material loss
8. Larger number of scheduled recipes
9. Reduced time to add new recipe
10. Reduced time to modify a recipe
11. Improved equipment utilization
12. Reduced downtime
13. Reduced number of manual operations
14. Lower engineering cost throughout the life cycle of your manufacturing process
15. Lower cost of capturing data, including less time to record batch data
16. Data availability is better: A larger quantity of higher-quality data is accessible for analysis

Various papers and articles have described the improvements S88 has made possible for many of these metrics. For example, it's not uncommon to see companies experiencing a 20 percent improvement in process cell throughput, a 20 percent reduction in batch cycle time, a 10 percent increase in product yield, a 33 percent reduction in project start-up time, or a 30 percent reduction in implementation cost because they used modular batch automation with S88. In recent years, the annual ISA Expo or ISA Tech events have had presentations about successful S88 projects. Visit www.isa.org to purchase conference proceedings. The World Batch Forum is also a great source for papers describing the benefits that can be derived from S88 implementations (visit www.wbf.org). You'll learn more about leveraging the knowledge of ISA and the World Batch Forum in Chapter 14.

Keep in mind that S88 was designed to handle all levels of automation. That is, S88 can be applied to a fully automated system or to a completely manual system ... or to anything in between.

If you plan to use original equipment manufacturers (OEMs), system integrators, or other vendors, make sure you hire only the ones that buy you the nicest dinners. Seriously, the S88 standard enables these firms to supply the appropriate tools for implementing batch control. But don't just *hire* a vendor to implement your system. That's like putting a coin in a slot machine, pulling the lever, and walking away. Stick around; see what you can win. You must embrace the standard yourself in order to get the complete payout.

To properly develop a successful batch control system, you need to define three important elements:

- How to make the product (recipes)
- What physical tools are needed to make the product (equipment)
- How to run that equipment (control activities)

You will read about each of these three elements in this book. We'll stop here for now to let this soak in and then move on to other basic concepts you should understand.

E-R Diagrams

Let's talk about E-R diagrams. While they have nothing to do with Michael Crichton's TV show, understanding this documentation technique might help you prevent some future hemorrhaging in your projects. Many books will place discussions about *entity-relationship* diagrams or other similar subjects in an appendix. We believe that understanding E-R diagrams is just too important a topic to defer, so we're going to present them right now.

E-R diagrams are a common way of describing objects in a system (entities) and how they correlate to one another (relationships). This type of documentation is very commonly used in the definition of data structures and flow, especially database architectures. Therefore, this technique has been very popular with IS/IT organizations.

There are four basic associations between objects in a system. In Figure 1.1, for each occurrence of A there is one and only one occurrence of B. The example E-R diagram reads "A unicycle has one and only one wheel."

Figure 1.1 One and Only One Association

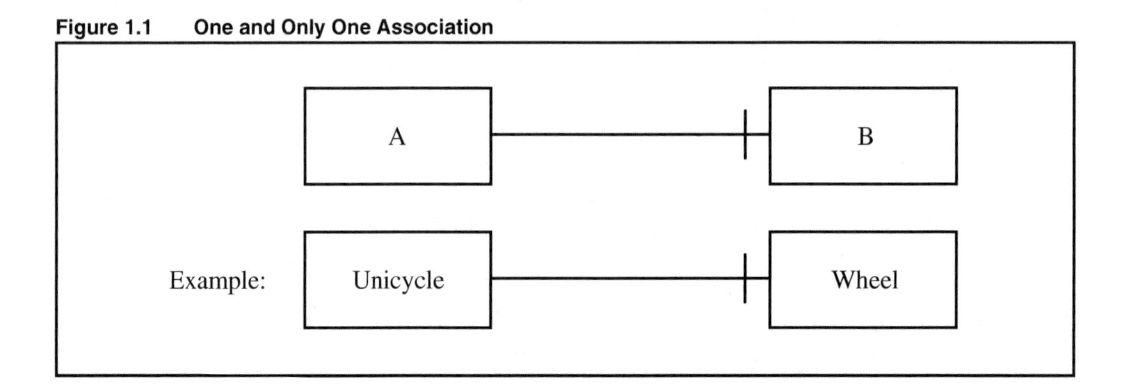

In Figure 1.2, for each occurrence of A there is zero or one occurrence of B. In this example, a car can have zero or one hatchbacks.

Figure 1.2 Zero or One Association

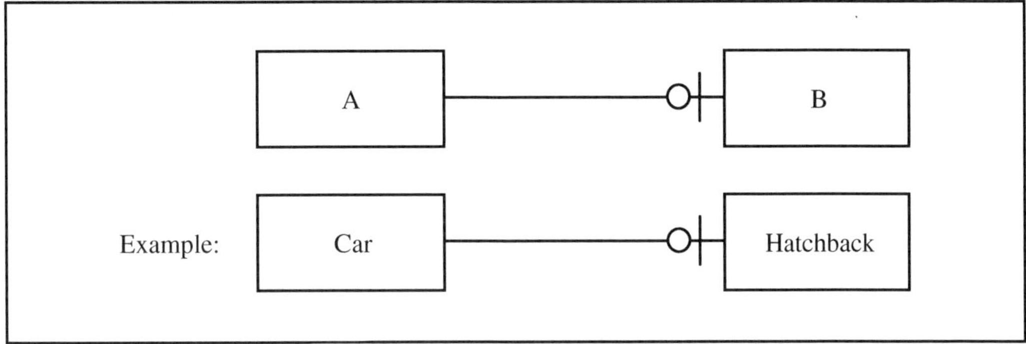

In Figure 1.3, for each occurrence of A there is one or more occurrences of B. In this example, a house has one or more bathrooms.

Figure 1.3 One or More Associations

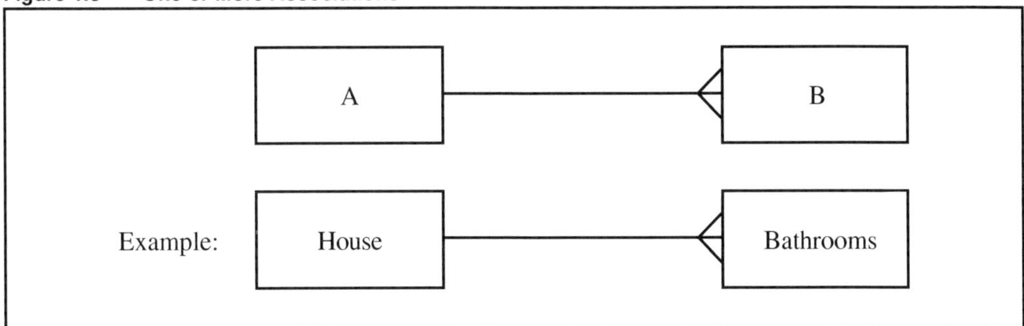

In Figure 1.4, for each occurrence of A there is zero, one, or more occurrences of B. A Ben & Jerry's employee gains zero, one, or more pounds during the first week on the job.

Figure 1.4 Zero, One, or More Associations

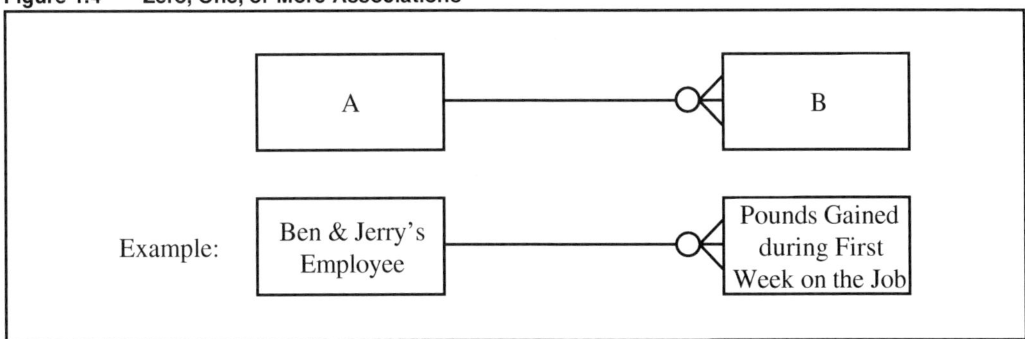

Okay, the example in Figure 1.4 is pushing it, but remember the phrase from college, "the freshman fifteen"? Well, at Ben & Jerry's they have "the Ben ten." It's hard enough to not gain weight by just working at an ice cream plant, but the company pushed us over the top with one of the best benefits around: three free pints a day, every day.

Anyway, many E-R diagrams include labeled associations, as in Figure 1.5. In the example, the association would be read as "A surfer dude owns one or more surfboards," or (more succinctly) "Surfer dude owns surfboards."

Figure 1.5 Labeled Association

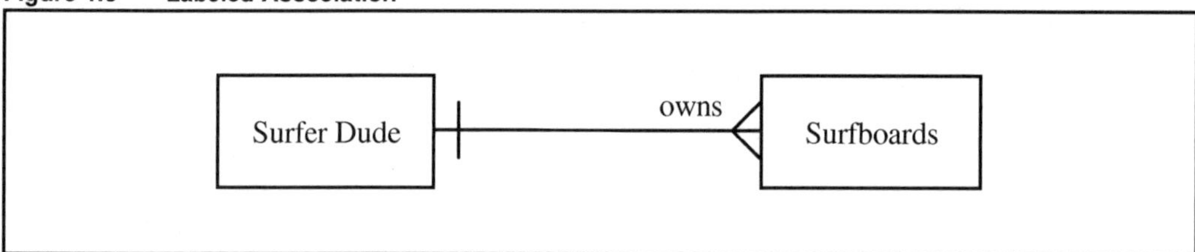

E-R diagrams need not focus on physical objects. In fact, many of the E-R diagrams we will be using in this book are more procedurally oriented. For example, in Figure 1.6, a procedure consists of an ordered set of operator instructions.

Figure 1.6 Procedural Association

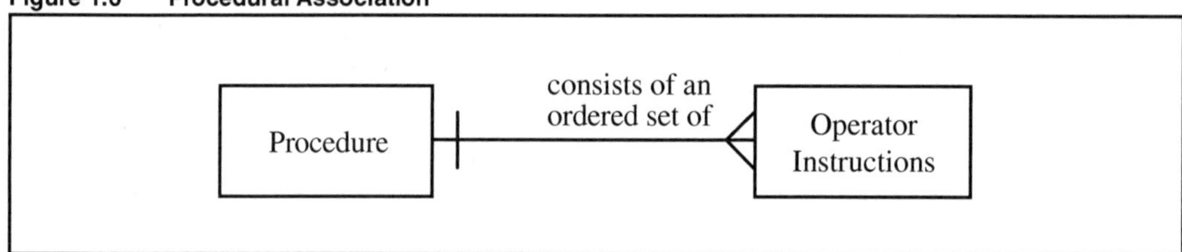

Enough of E-R diagrams.

Sequential Function Charts

The International Electrotechnical Commission standard IEC-61131-3 includes something called a sequential function chart (SFC) as a programming language. It is also common to use this nomenclature as a documentation framework. We've illustrated an SFC in Figure 1.7. Existing products, including PLC programming software and batch control packages, use SFCs to represent executable procedures. In this section, we're first going to introduce two common elements of an SFC: *steps* and *transitions*.

A block containing either a *step number* or *step name* represents an SFC step. Vertical lines link the steps. The initial step of a sequence is drawn as a box with a double line. A step is either active or inactive at any given time. We'll see in a little bit how more than one step may be active at a time.

A horizontal line represents a *transition* across the vertical link between two steps. It lists the condition for transferring control from the *active* step preceding the transition to the step following the transition. That is, when a transition is true, or *fires*, the active step immediately before the transition becomes inactive and the step immediately after the transition becomes active. We talk about an active step because a transition can be true anytime, but for a transition to activate the step following it the step preceding it must be active when the transition fires. Figure 1.7 shows the step and transition elements of a sequential function chart, as well as the sequence terminator. (Note that IEC-61131-3 does not define a sequence terminator. That standard requires that an SFC end with a step and not a transition. Some batch control software packages, including OpenBatch, use this terminator.)

Figure 1.7 Elements of a Sequential Function Chart

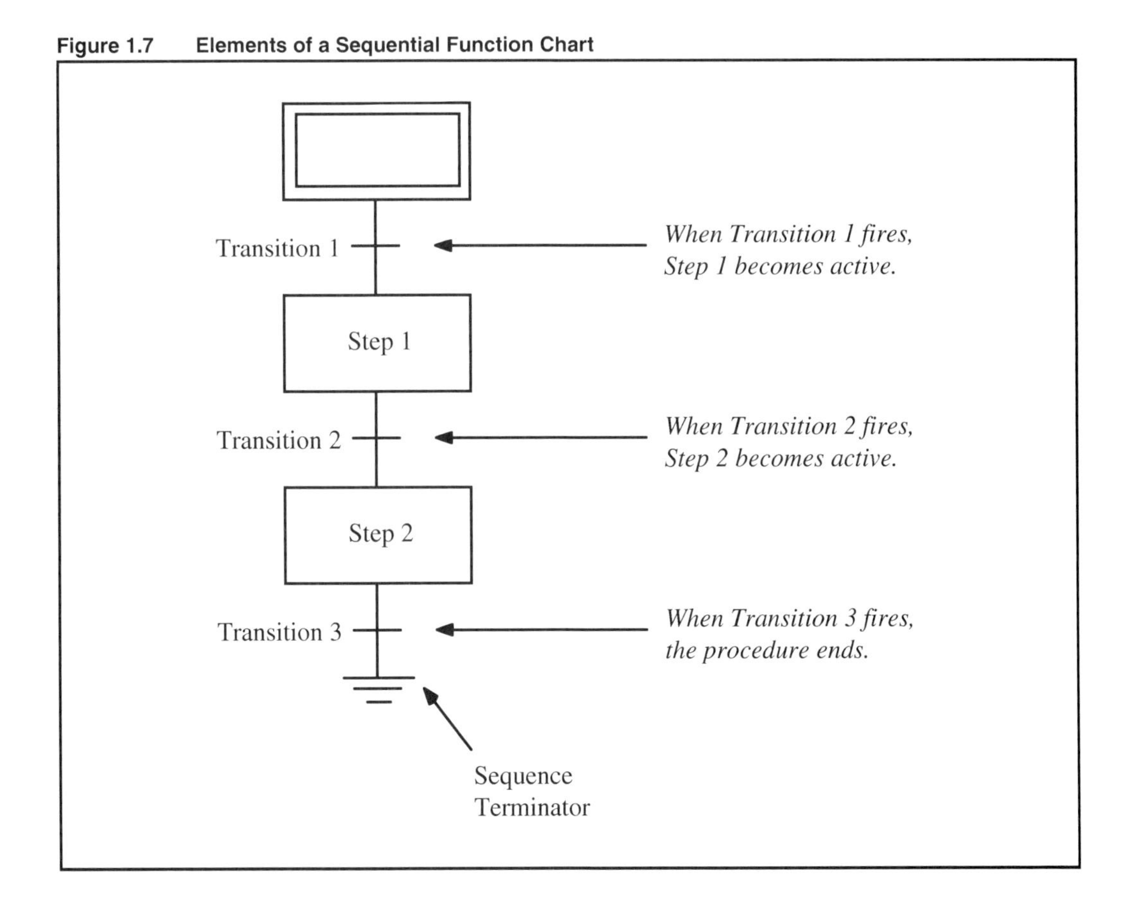

Figure 1.8 adds some real-world features to our example.

Figure 1.8 SFC Example: Mixing Milk and Cream

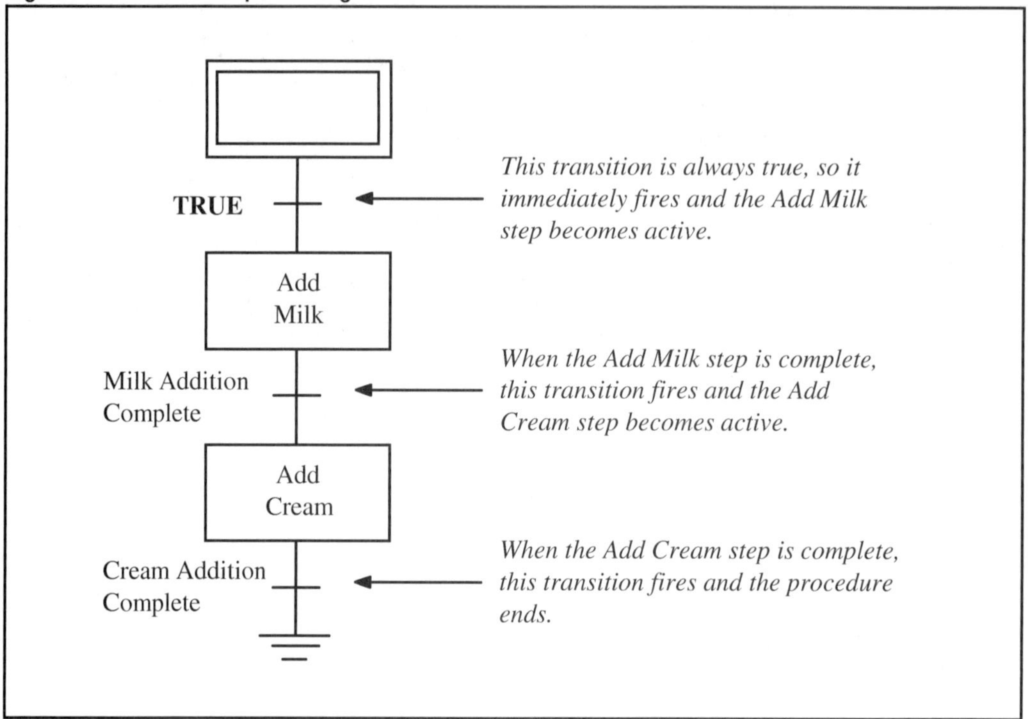

Notice a couple things in Figure 1.8. First, the "initial step" is not labeled; it is just a double box. Second, a **TRUE** transition is allowed. Keep in mind that a **TRUE** transition will immediately pass control to the next step or steps in line. If a **TRUE** transition existed between *Add Milk* and *Add Cream*, *Add Milk* would stop and *Add Cream* would start right away, even though not all the milk had been added. Remember that as long as the preceding step is active and the transition is true, the preceding step stops—even if it is not yet complete—and the following step starts. To prevent this from happening, you'll find the most common type of transition is the "step complete," as shown in Figure 1.8.

SFCs handle *AND/OR* branching pretty well also. An *OR* condition, or what IEC-61131-3 calls a *divergence of sequence selection*, is shown in Figure 1.9.

In the example in Figure 1.9, we only add one type of cocoa, high-fat or low-fat, depending on what the recipe calls out. Figure 1.10 builds on Figure 1.9 to show how the sequence converges. Note how only one alternate path is taken, but no matter which cocoa is used the sequence converges back to adding cream. When designing your control scheme, it's always convenient to construct mutually exclusive conditions in your transitions. Even if both transitions were to fire above (let's say we mixed both types of cocoa in a recipe), SFC rules only allow one path to run. Generally, we like to think the "left" path has the priority, but that of course depends on the SFC implementation of the product you're using. So, to be sure, check with the product support group to see how its rules work.

Figure 1.9 Divergence of Sequence Selection (OR Condition)

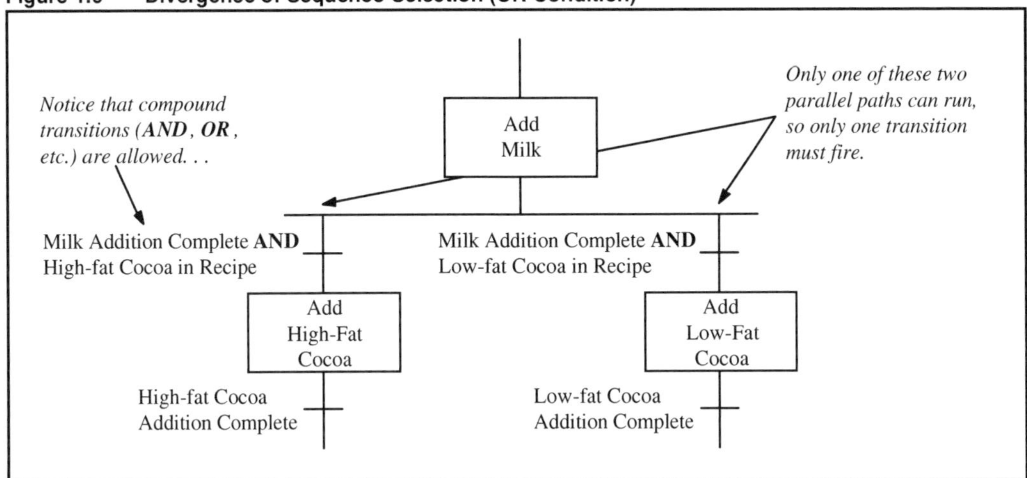

Figure 1.10 Convergence of Sequence Selection

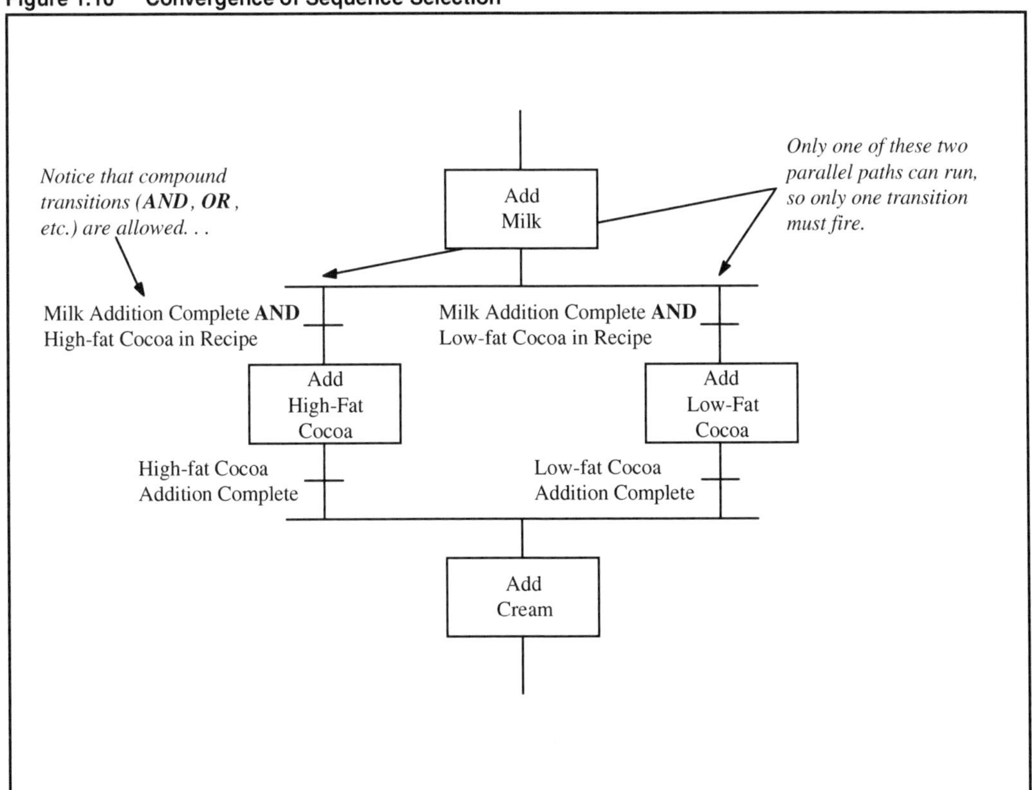

Figure 1.11 shows an *AND* condition, or a *simultaneous divergence.*

Figure 1.11 Simultaneous Divergence (AND Condition)

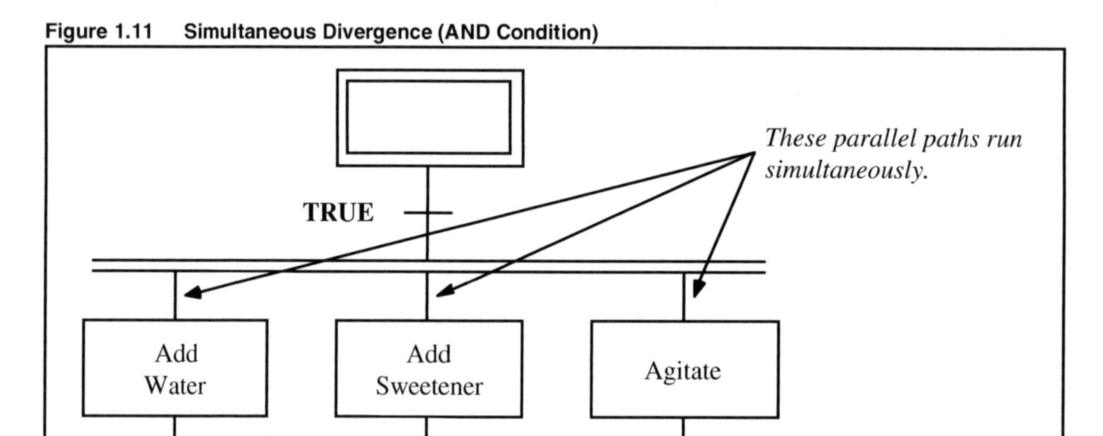

Notice how a single transition (in this case, *TRUE*) starts all three steps simultaneously. The double horizontal line is called the *line of synchronization.* With a simultaneous divergence, the transition must occur above the line of synchronization. Figure 1.12 shows the corresponding convergence to this condition.

Figure 1.12 Simultaneous Convergence

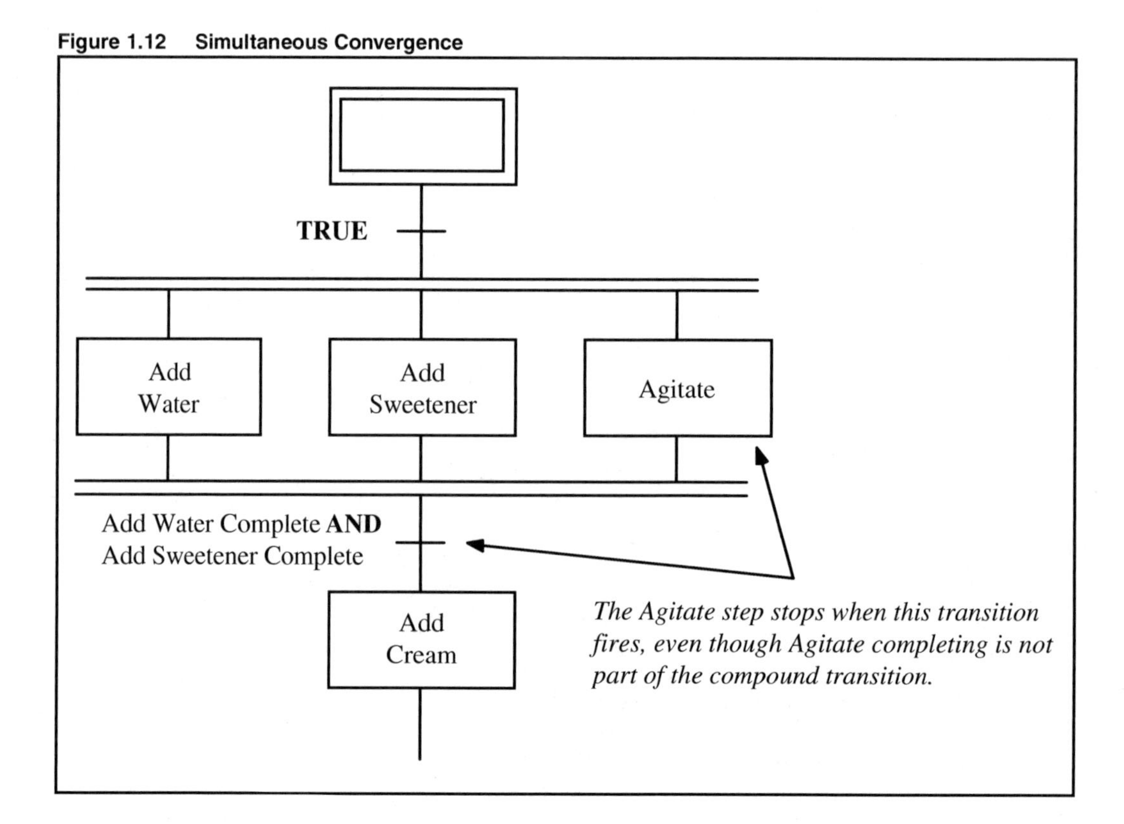

Notice that a compound transition is allowed (*Add Water* completes *AND Add Sweetener* completes). Also pay attention to what happens after that transition fires. According to SFC rules, when this transition is true *Add Cream* becomes active, and *Add Water, Add Sweetener,* and *Agitate* all become inactive. So with *Agitate,* you now see how a step can become inactive even without a transition indicating that the step has completed.

In at least some batch control products, you can skip steps as well, which is shown in Figure 1.13.

Figure 1.13 Skipping Steps

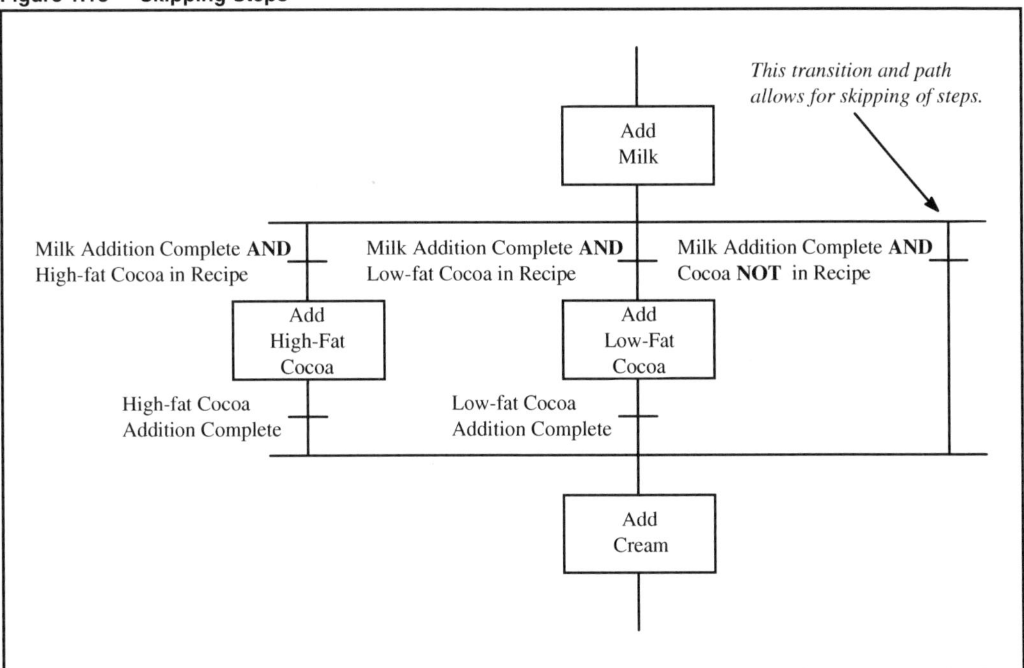

One last rule on SFCs. If you look through all of the preceding diagrams, you'll notice that no two steps occur without a transition in between, and no two transitions occur without a step in between. This is a golden rule with SFCs. If two transitions need to fire in order for a step to start, perhaps you can create a *compound transition.*

We're tired of talking about SFCs, so let's talk about mix-making.

A Typical Mix-Making System

Figure 1.14 is a diagram of a typical mix-making system. While it is not very "engineering-ish," it'll have to do.

Figure 1.14 A Typical Mix-making Process

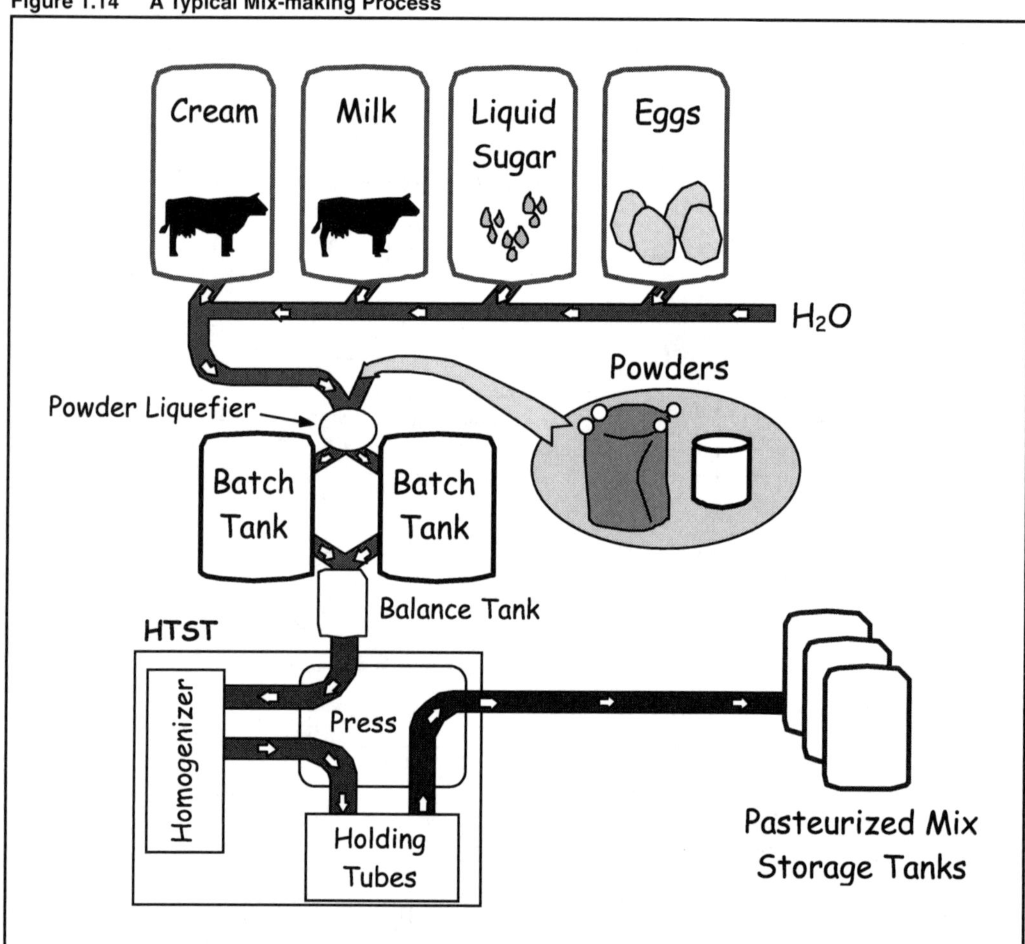

All the examples in this book will be concerned only with the mix-making batching process. In Figure 1.14, this process is represented by all the equipment up to and including the two batch tanks. Most ice cream companies make use of both liquid and powder ingredients. All liquid ingredients, including cream, milk, liquid sugars, eggs, and water can be dispensed using flowmeters or load cells. All powder ingredients, including powdered sugars and cocoa, are generally dispensed using load cells.

To make mix, you simply combine the necessary ingredients according to a recipe so as to form a mix "soup," which is not very different from making ice cream in a freezer at home. But while at home you may make a half gallon or gallon of ice cream mix, ice cream companies sometimes make thousands of gallons of mix *per batch*. Depending on the mix type, each batch may have over a thousand gallons of cream or milk and may have over a thousand pounds of cocoa. The batch tanks generally operate in parallel, though not simultaneously: mix is made in one, then the other. While one tank is batching, the other is feeding product to a pasteurization process.

Often product types and production scheduling require a high level of process flexibility. It's not uncommon for a company to have dozens of different mix recipes and to run several different recipes in one day. Mix makers try to do this without shutting down the pasteurizer. To complicate matters further, the amount of fat and nonfat solids present in milk can vary with each truckload. The amount of solids in milk depends on the cow's diet and on the season. Some companies order standardized cream and milk, which results in the same solids potencies for every truckload. Other companies don't, instead varying their recipe ingredient quantities to achieve consistent final mix specifications. If tomorrow's load of cream has less fat than today's, operators will use more gallons of cream but may balance that with fewer gallons of milk or water. (Cream is sold by pounds of fat, not gallons, so the cost works out to be about the same regardless of potency.) Therefore, even though a plant may be running the same mix type, once operators switch to a different raw dairy tank the recipe quantities may change.

Did you get all that? It doesn't really matter; we'll repeat the important points as necessary. Before we go any further, the next two chapters talk about some necessary business and project management topics. Stay tuned; this is good information.

Pasteurizing Ice Cream Mix

We thought you might like us to explain a popular method for pasteurizing ice cream mix, even though understanding a pasteurization process isn't vital to understanding S88. Downstream from the batch tanks is a high-temperature short-time (HTST) pasteurization process. (Old dairymen sometimes call this process "the short time.") Pasteurization is based on a time-temperature curve. The higher the temperature, the less time the mix needs to be held at that temperature to be considered pasteurized. HTST involves somewhat high temperatures (175-200°F) and somewhat short holding times (20-50 seconds).

Follow along with Figure 1.15, which shows a more detailed view of the HTST pasteurization process, while we explain the steps:

1. **As mix leaves one of the two batch tanks, it travels to the hundred gallon or so balance tank, which serves to ensure that the process remains charged with mix at all times.**

(continued)

(Pasteurizing Ice Cream Mix continued)

2. From the balance tank, the mix is pumped to the first of three sections of the pasteurizer. In the diagram, the pasteurizer is labeled "Press." *Press* is slang for the pasteurizer, as it is merely a three-section plate heat exchanger with the plates pressed tightly to one another. It sort of looks like a giant automobile radiator. In the first press section–the regeneration section, or "regen"–the mix is preheated to around 150-170°F.
3. At this temperature, the mix passes through the homogenizer. The homogenizer pulverizes the fat globules into microscopic-sized particles. (The smaller and more equally distributed the fat globules are, the smoother the ice cream tastes.)
4. After being homogenized, the preheated mix moves to the heating section of the press, where its temperature rises to 176-200°F.
5. The mix then travels through a snaking set of holding tubes, which provide a physical means for ensuring that the mix is held at the pasteurization temperature for the required period of time. At the end of the holding tubes is a temperature transmitter and a divert valve.
6. If the temperature of the mix at the end of the holding tubes is not at or above the minimum acceptable pasteurization temperature, the divert valve will force the mix to the balance tank, where it is recycled back through the HTST process.

   The HTST system is also pressurized, so that product leaving the pasteurizer is at a higher pressure than product entering the unit. If a leak occurs, pasteurized product will flow to the raw part of the process, not the other way around. As an added safeguard, if a minimum pressure differential is not maintained, the divert valve will force the mix to the balance tank.
7. If the pasteurization process is in normal "forward flow" mode, the mix travels through the "back side" of the regen section to have its heat transferred to the mix coming from the balance tank. After regen, the mix enters the final section of the press–cooling–where it is chilled to about 35-40°F for storage.

   The "divert valves" are controlled by a separate control system, often an independent programmable logic controller. This PLC is enclosed in a separate panel and is sealed by the state regulatory authority. If a plant makes any alterations to this control system, a state regulatory agency generally must reinspect the system. Information about process and equipment states is generally shared between the two PLCs via hardwired interlocks.

Figure 1.15 The Pasteurization Process

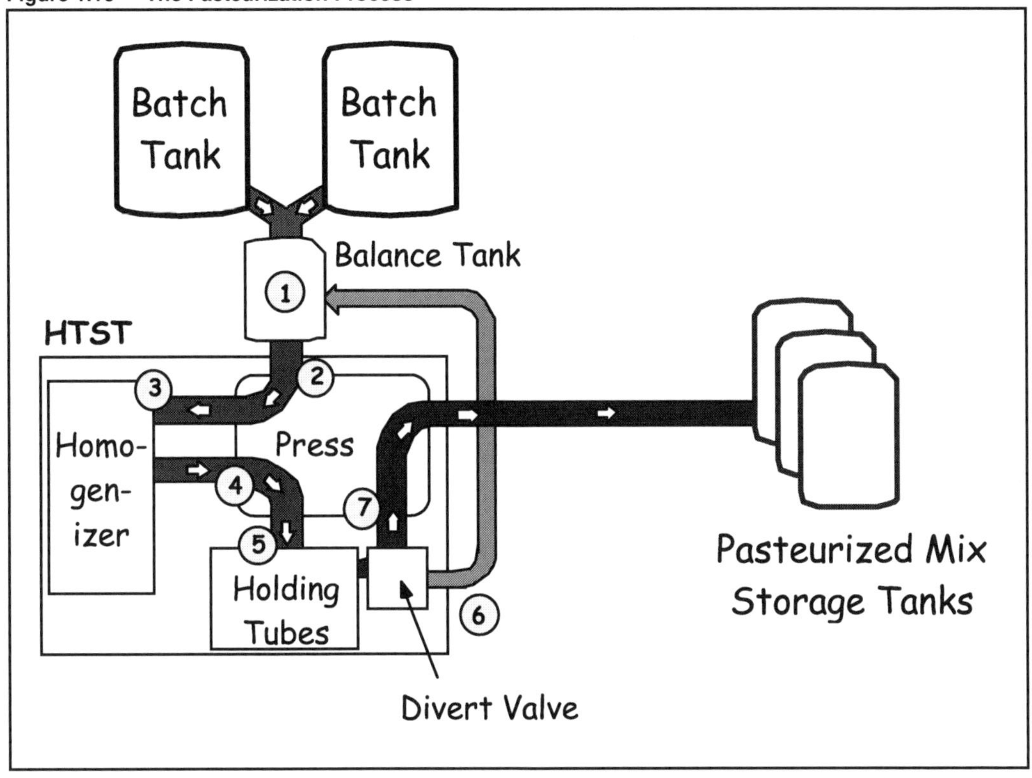

2

Are You Ready to Go Yet?

"This chapter is all about begging for permission to start your project. (Sorry, we don't have a chapter on begging for forgiveness for completing a project that you never told anyone about.) If you already have the go-ahead, congratulations, you lucky dog. You're the envy of the remaining 99 percent of us. Now quit wasting your time and skip ahead to Chapter 3. For the unlucky ones left reading this chapter, we're going to talk about selling management on the need for an S88 supervisory system.

Gathering Requirements

How many times have you finished a project only to find that you could have done so much better if you'd known just a little more about what was required? Maybe you had to redesign your solution halfway through the project, or maybe your customers only returned blank stares when you exclaimed, "Ta-daaa!"

Hey, let's face it, requirements gathering doesn't hold the distinction of being the most glamorous part of a project. Many engineers either want to start designing or start building. But let's not just point the finger at engineers. You managers sometimes don't help much either. You debate solutions, ponder, wish, and otherwise delay approving the project for so long that by the time the project proposal has the necessary signatures, the original proposed schedule has slipped by weeks or months. You somehow forget why the schedule slipped, mumble something like "How could we let this happen?" and insist that the project be finished by the original date. This can often result in rushed requirements gathering.

Regardless of the reason, many companies do not always take sufficient time to analyze requirements. Without clear requirements, project "costs" will rise—and not just capital costs. Don't forget about rising operational costs, schedule slips, inadequate consideration of safety and quality issues, regulatory shortcomings, and a poor design that simply does not meet customer requirements.

Depending on your company's policies, you may gather and analyze project requirements before or after you submit the project proposal. In justifying our mix-making supervisory system at Ben & Jerry's, we chose to gather requirements *before* submitting the proposal. We attached the requirements document to the proposal to help substantiate our analysis of company needs.

It's important that you gather requirements; *how* you go about it isn't critical. We have found that a quick, surgical thrust works best for us. (If that fails, we try a sledgehammer approach.) For "smaller" projects (less than $100,000 in capital costs), the lead engineer typically interviews key people within a day or two, including people from operations, maintenance, quality assurance, and finance. The lead asks another engineer for feedback on a preliminary requirements document, then issues the first draft. For larger projects, we tend to include more people—sometimes from other sites—iterate an extra draft or two, and host a review. If we've made assumptions about the project, we state them in the requirements document, and we also include unanswered questions. Sometimes one of the reviewers can answer a question or two.

Carefully keep in mind the political benefits of gathering requirements. First, it gives you another chance to sell the project to people who are not yet convinced of the solution. A completed requirements document clearly outlines the expected benefits of the solution and may present concepts in different terms than before. Second, gathering requirements gives you a great opportunity to ensure that your customers assume ownership of the project. By including all the right people in the requirements-gathering process, their needs, thoughts, and fears will be recognized and addressed. But the most important thing about including all the right people is quite simple: if the project fails, the blame is spread among more people. At the very worst, you don't go down alone.

Our requirements gathering at Ben & Jerry's included assessing our need for an S88-aware batch control system and for other supervisory system components, such as a database and a human-machine interface (HMI). (We safely made an assumption at this point that we needed an S88-based solution.) A requirements document might include the following five sections:

1. Document purpose
2. System objectives
3. Project scope
4. Implementation plan
5. Tools required

The first section, the requirements document's purpose, can be as brief or as elaborate as your audience requires. For the second section, on system objectives, here are some typical objectives:

> *Better understand where variances occur in the process*—Understanding variances is expected to lead to the identification of waste and a more consistent product, subsequently reducing operational costs.
>
> *Reduce system troubleshooting time*—Improving operators' knowledge of the system and ensuring quicker reporting when a process is out of control will result in reduced downtime and reduced risk for loss of quality. (Sounds good, doesn't it? Who were we kidding? The most important

thing for us was to reduce the number of 2:00 a.m. calls we received at home when the system didn't work!)

Reduce unnecessary and tedious manual calculations and paperwork—Less manual work frees operators for more important tasks and reduces associated errors (e.g., transcription, keypunch).

Better understand ingredient usage—The supervisory system will help operations and finance better understand the reasons for inventory adjustments (i.e., the reasons for manufacturing variances or material waste).

Improve flexibility of mix-making system—Making new mix recipes, using new ingredients, and modifying CIP procedures is expected to be less cumbersome. (CIP is "Clean-In-Place," a method for automatically cleaning equipment without tearing it apart.) The responsibility for making minor system changes for new recipes and ingredients will shift from engineering to operations.

The first and fourth objectives can be tied to the proposed system in its role as an "enabler." Much of the time, plant floor automation serves as an enabler, helping to identify where problems are occurring. Without the automation these problems may never be identified.

Like our requirements document for Ben & Jerry's, section 3 of your requirements document, on project scope, might list specific tasks that the proposed system will perform:

Automatically collect accurate data on the amount of raw ingredients received, raw ingredients used in each batch, mix made, and mix transferred to the flavor vats. This is expected not only to reduce tedious paperwork and associated errors but also provide additional information that is currently too difficult to collect.

Automatically calculate variances and where they occur. This is determined by analyzing differences in the data collected in the preceding automatic collection task.

Improve batch operations and calculations. New software is expected to help tighten batch tolerances; to provide greater flexibility for new products, ingredients, and process changes; to reduce paperwork by automatically calculating recipes; and to provide better batch result information.

Provide a detailed explanation to operators as to why a function has failed or will not start. Providing diagnostic capabilities is expected to reduce downtime and loss of quality.

Section 4 of the requirements document, on the implementation plan, should provide further detail on each item listed in the project scope section. Finally, section 5, on the tools required, can simply list the hardware and software needed to accomplish the objectives and scope listed in the requirements document.

Selling the Concept (Getting Funding)

As much as you may wish it were otherwise, writing an S88 proposal is no different than writing any capital project proposal: formal justification is key to obtaining funding approval. Each company treats automation justifications differently, and some understand the value of flexibility and enabling technology better than others. At Ben & Jerry's, we chose to base our proposal on the potential for cost savings that the supervisory system would provide. Keep in mind that this early stage of your project can be hampered by fear and politics. People will be:

- Afraid of new technology (most likely because they don't understand it)
- Trying to get the new technology at their site first, not yours
- Attempting to wrest control of the new technology away from your department

In our careers, we have experienced all three situations. Let us be the first to assure you that those early days seeking formal justification were some of the happiest and most blissful of our lives—Not. Look, do whatever you need to do to avoid these pitfalls: educate, build consensus, whatever. We're not here to advise you politically. (Larry will tell you that's the *last* kind of advice you want from Jim.) Take it from us, physically visible tools, like HMIs, attract a lot of attention. If choosing a specific HMI is going to create issues, make sure you propose an S88 solution that is flexible enough to work with different HMI products.

If your project can be easily justified in terms of cost savings, a return on investment (ROI)-based proposal may be the way to go. If you are anticipating cost savings but cannot quantify the details, you may want to consider discussing additional benefits like the following:

- *Quality issues*—Automation may help increase the consistency and quality of products. Don't neglect to tell management that consistency and quality of *information* is important too. (We believe that information is the second-most important asset of any organization.)
- *Responsiveness and flexibility*—The capability to adjust rapidly, efficiently, and effectively to required changes creates a distinct advantage in successfully and quickly bringing products to the marketplace.
- *Automation as an "enabler"*—Supervisory systems can serve as "enablers" that allow operators and managers to better understand where variances and waste occur. With this information, production processes and procedures can be altered to improve quality, minimize waste, and reduce costs. Without automation these problems may never even be identified.

A successful proposal clearly shows that the "seller" understands the business's problems. It should show that the proposed project is part of a total vision and that it supports the company's manufacturing strategy or strategic goals. The proposal should make good use of visualization techniques, such as charts and graphics, where appropriate and should discuss case or benchmark studies. Of course, if necessary, don't be afraid to "package" the proposal to the bias of the "buyer." Clearly match the project's benefits to the personal motivators of those with the magic signing pen.

Many companies have a standard format for project proposals, so we're not going to outline that here.

If you've already created a requirements document, heed this advice:

- In the project detail section of your proposal, refer back to your requirements document, or even insert pertinent sections from it. This will help show management that you're not making up your claims.
- Unless you enjoy the excruciating pain of approval delays or rewriting proposals, map the project benefits listed in the benefits section of the project proposal to the system objectives outlined in your requirements document, clearly showing how your project will satisfy plant needs.

When listing benefits, use quantifiable facts. If you have financial data, use it, but employ tables and graphs so management can visualize it. Do the same when discussing the time savings derived from reducing the manual paperwork activities of operators. If your company's goal is to find better uses for operators' time, not to reduce head count, you may not want to convert the paperwork hours saved into dollars. Don't forget to include time estimates for correcting errors associated with manual data entry. There is also value in timeliness: data collected automatically in real time has a higher value than data entered manually at the end of the day.

You may also wish to discuss the project benefit of enhancing system flexibility. To quantify this benefit, consider the amount of engineering time that was required to design, implement, and test ladder logic changes during a recent project. Then explain how this time could be cut if an S88 batch management solution were installed. If you can get it, don't forget to include information that shows how enhanced system flexibility will allow products to enter the market faster.

Enhancing product quality is a definite benefit. Investigate whether increased product quality will result when operators have more detailed, *real-time* information about the batching processes. This real-time information can provide warnings to operators if processing parameters are expected to fall outside of specification. The operators can then correct the problem before the product falls out of spec.

Good luck with your proposal.

Starting (What You Hope Will Be) a Successful Project

Now we're going to talk about planning and preparing for your S88 implementation. This chapter mostly contains project management stuff, but not a lot of it. After all, this isn't a book about project management; it's a book about implementing S88. For your enjoyment, and at no extra charge, we have included some important things that we learned about project management during our Ben & Jerry's project and others.

Step One of a Successful Project

Buy your operators a new desk. They'll be your best friends throughout the project.

All right, so maybe you don't really need to buy a desk—not a big one, anyway. But your operators are among your "customers," and you need to make sure your customers are happy. Without your customers' support, your project is doomed. Getting your management's support is absolutely necessary, but that support really only gets you time and money. For your project to succeed, you must satisfy the needs and wishes of the people who will actually be using the system. If you don't design and implement a solution that satisfies your customer, that customer isn't going to use your whiz-bang system. If no one uses your system, it's a failure.

With luck, when you were gathering requirements you focused on the visible needs of your customer. Sometimes this is simple, sometimes not. If your customer is very aware of your process and its potential, he or she will have all kinds of ideas on how to improve it. But in many cases, you may have to dig a little to find out how to improve your customer's work life. And, by George, that's what your job is really all about: *Improving your customer's work life!*

The operator was the center of manufacturing at Ben & Jerry's. As engineers, we didn't simply try to make a piece of equipment or system run better; our mission was to focus on the needs of our customers—the operators—in order to make their work lives better. Likewise, we worked hard to ensure that we didn't have just "button pushers"; we wanted educated customers!

When operators can do their jobs better with fewer obstacles, they make consistently great products at the lowest possible cost and highest possible quality. When that happens, everyone benefits: the operator, the engineer, the CEO, shareholders, even the little kid next door eating some Chocolate Chip Cookie Dough ice cream.

While we were at Ben & Jerry's, we made things better for our customers by implementing solutions that reduced tedious paperwork, replaced laborious manual tasks with automated counterparts, improved manufacturing efficiencies, and added product and ingredient flexibility to the process. This sounds great, but don't forget about realities. Hey, even Ben & Jerry's had to consider reality. The company's products were in such demand, especially during the summer, that the plants often ran six days a week. As time can be a luxury in any project, this schedule presented problems. One down day a week just wasn't good enough for many of our projects. Having an S88-aware batching system helped Ben & Jerry's roll out new products and make changes to processes in less time, reducing the need for weekend work.

Getting back to the desk, we actually bought it because we needed space for a supervisory system desktop computer. We knew we didn't have to buy it to get our mix makers' support. The mix makers we worked with were awesome and any engineer's dream customers. The plant's lead mix maker is a Ben & Jerry's old-timer, who had started with the company back when steam engines turned agitators. She knows the mix-making process better than anyone (including us).

All three mix makers at the St. Albans plant relied on automation, but they would not allow themselves to become dependent upon it. They realized they needed it to help them make mix better, but they used it as the tool it was meant to be. That way, if automation failed they wouldn't throw in the towel. Instead, they'd find ways of working around the problem until an engineer had a chance to correct it. Working with people like this was critical to the success of the project.

All your customers may not be operators. Our plant finance group was instrumental in selling and encouraging the S88 project. The process and inventory information provided by the supervisory systems helped them keep their records more current. The plant cost accountant helped us locate sources of manufacturing variances by evaluating process data. He could do his job better when the systems provided deeper and more accurate data.

We included our lead mix maker and plant cost accountant in all aspects of the project right from the beginning. Both helped us determine the project's requirements by clearly explaining what they needed to make their jobs easier. We got critical data about the mix-making process and learned what data could help better identify areas of weakness in mix-making. In the end, the lead mix maker flew with Jim to Phoenix for the first round of OpenBatch training. (The class was held in January; their arms really had to be twisted.)

Moving Forward with a Successful Project

First of all, if you didn't do so before submitting your proposal, complete the formal requirements-gathering process. As we mentioned in Chapter 2, be sure to involve the right people during this process to help obtain "ownership" and support for the project. We were lucky. Our lead mix maker was excited enough and understood enough about this project that ownership was never an issue with her.

Once you've completed documenting your requirements, host a formal kickoff meeting. Invite operators, managers, engineers, or anyone else you believe will feel the impact of or have an impact on the project. Don't forget to include people from other sites, if they can help. At the meeting, review the project requirements, scope, schedule, and costs. Allow everyone to freely ask questions to help eliminate any last-minute fears or uncertainties, but keep the meeting focused on the right issues. For example, if your audience is not overly technical, keep steering the discussion to the business issues the system will address.

Be careful not to severely underestimate the amount of downtime needed for installation and testing or the amount of raw material that may be wasted during start-up. It's been our experience that while managers may not like the terms *downtime* or *waste*, these issues can be negotiated. Keep an open and creative mind. The St. Albans production manager was another Ben & Jerry's old-timer, about fifth highest on the company seniority list (including Ben and Jerry!). She was very fair and had a grasp of both details and the big picture. One of our standing goals was to never, ever disappoint or surprise her.

If you're planning to use a system integrator, an OEM, or a consulting firm to help design or implement all or a part of your system, here are a few important things we have learned on several projects:

- *Include requirements for system acceptance in contracts*—Most contracts will include standard items like cost, schedule, and performance requirements. Many contracts don't include your expectations for system acceptance and particularly the level of validation testing you desire. See Chapter 13 for more information on validation test plans.
- *Work hard to ensure compliance with published standards and system specifications*—Most contracts will require adherence to standards, including mechanical, electrical, and software specs. (Don't forget S88!) However, many companies assume that the vendor performing the work is following the requirements, and they won't bother to perform periodic

audits. We've been burned on this before. It's not that our vendors were being deceptive or trying to save a buck—most of the time the spec wasn't followed because of a misunderstanding or an honest omission. Hosting periodic reviews with your vendor can save you from some nasty headaches later. Sure, you can hold the vendor accountable when you find discrepancies at the conclusion of the project. But then you'll face a choice: have the vendor correct the discrepancies, meaning the project will probably finish late, or have the project finish on time without meeting all the specs. When the vendor leaves, you'll be the engineer with a late or incomplete project on your hands.

- *You're not hiring a firm, you're hiring skills*—Interview the individuals who will be on your site. Use the results from those interviews to help determine which vendor, integrator, or OEM to hire. Obviously, technical and project skills are needed, but don't overlook interpersonal skills. Ben & Jerry's had its own special way of handling projects. Sometimes it took the right type of person to thrive there. When negotiating a contract with a firm, consider including a requirement that specific employees of the firm work with you.
- *Consultants, vendors, and OEMs aren't perfect*—You hired them for a reason. Perhaps you needed expertise or you were short on internal resources. In either case, you should be able to expect your contracted help to know what they're talking about. However, question their recommendations and estimates as much as you would those from any other source. It's always frustrating to find that a simple miscalculation or error has created havoc. And just because you've hired a vendor to help you doesn't mean you should assume that the vendor is *the* expert. We're probably going to catch a lot of flak from some of these companies for saying this, but it may not be in your best interest to hire a firm whose first S88 project is yours.

Don't wait until the project is complete to develop training manuals or programs. Start them soon after the design is complete. This helps in two ways. First, if you start right away your manuals will be ready by the time you want to train your users. Second, writing the manuals may provide you with some additional insight from the operator's point of view. You may catch things that you want to change before it's too late.

Fix software bugs *as you find them* while testing your system. (And, yes, you *will* find them.) This is smart for several reasons:

- Fixing bugs as you go provides damage control. The earlier you discover your mistakes, the less likely you are to repeat them.
- By keeping the bug count near zero, you'll be in a better position to track your progress against the project schedule. Instead of trying to guess how long it will take to finish 16 software features and 102 bug fixes, you just have to guess how long it will take to finish the 16 features.

- As projects near completion, people want to finish them. Programmers may tend to blow off the last few bug fixes, thinking they'll get around to fixing them later.
- Telling management that all software features are in and that your programmers are working on nothing but bug fixes just doesn't sound good.
- Bugs are a form of negative feedback that keep fast but less-than-thorough programmers in check. If programmers aren't allowed to start working on new features until they have fixed all the bugs on the old ones, they're prevented from spreading half-implemented features throughout the project.
- When the system goes on line for the first time, bug-free or not, you'll have less freedom to make changes. The fewer bugs at the beginning, the less pressure you'll be under.

Finally, pay fantastic attention to detail, especially with your operator interfaces. You can have the best code in the world, but if your HMI is sloppy, you look sloppy. No detail is too picky. Align those fields perfectly! Keep colors consistent. Don't overuse flashing text or graphics. Don't create clutter just to fit everything on one screen or form. And, for Pete's sake, no spelling errors!

Enough of planning, proposing, and managing. Ready to really start learning about S88? Great, but please remember as we begin the ride: there can be no eating, drinking, smoking, or the use of flash photography…

4

THE PHYSICAL MODEL

S88.01 defines several models that describe the equipment and procedure hierarchies necessary to make batches. In this chapter and the next, we will introduce and discuss these models in detail. We're going to start with the *physical model* because it probably is the model that is easiest to grasp. This model is used to describe the physical equipment associated with your batching operations. The model, shown in Figure 4.1, has seven levels. Let's start describing the sections from the top.

ENTERPRISE AND SITE LEVELS

Enterprise is really a fancy name for *company.* Some consulting firm decided years ago that *enterprise* sounded more sophisticated or more businesslike than *company.* It stuck. Now there are terms like *enterprise resource planning* (ERP) systems.

In very large companies, the "enterprise" might refer to a division or business unit. But as far as we're concerned, *enterprise* refers to a company. In our case, it was Ben & Jerry's. The S88 standard states that corporate, divisional, or business-unit activities and decisions are made at the enterprise level. This includes deciding which products will be made, where they will be made, and when they will be made.

Likewise, *site* is another name for *plant*. Geography is probably the most common method for determining what a site is, but that doesn't mean two sites cannot be physically adjacent. Different sites can make different products, but they don't have to. Likewise, different sites can have different manufacturing processes, but they may be the same.

Figure 4.2 is how we viewed the Ben & Jerry's enterprise and its three Vermont sites, Springfield, St. Albans, and Waterbury. Since S88 is concerned with batch control, the only applicable sites were the three manufacturing plants, even though the company also had a corporate office and a distribution center. Waterbury, the oldest of the plants, packaged ice cream in only pint-sized containers. Springfield produced novelties, bulk tubs for retail "scoop shops," and ice cream in quart-sized containers. Springfield could also produce pints but was not normally scheduled to do so. St. Albans, the newest and largest plant, produced pints and 140 ml, single-serving cups. Although pints could be produced at any site, the process equipment was slightly different at each.

Figure 4.1 S88.01 Physical Model

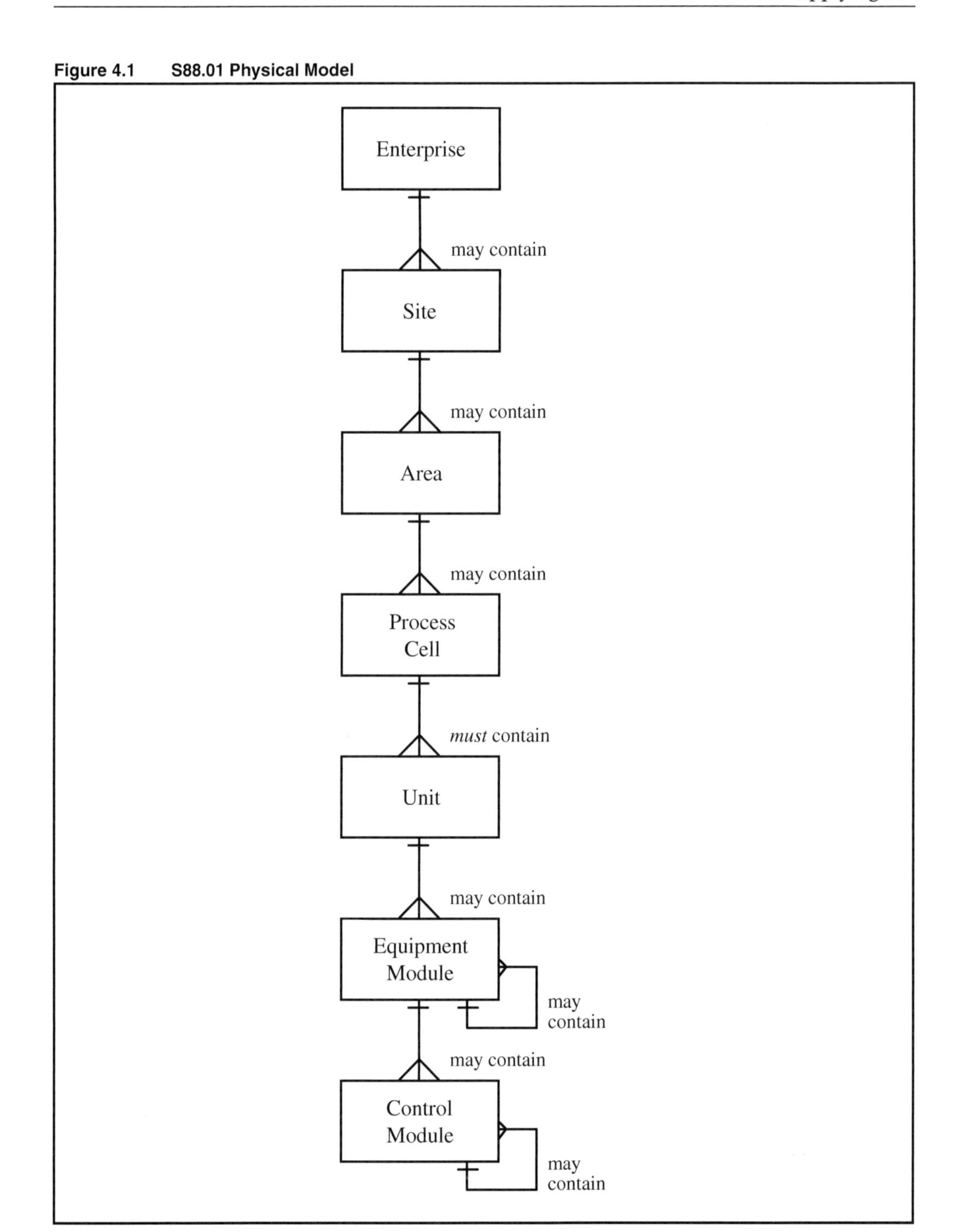

Figure 4.2 Ben & Jerry's Enterprise and Sites

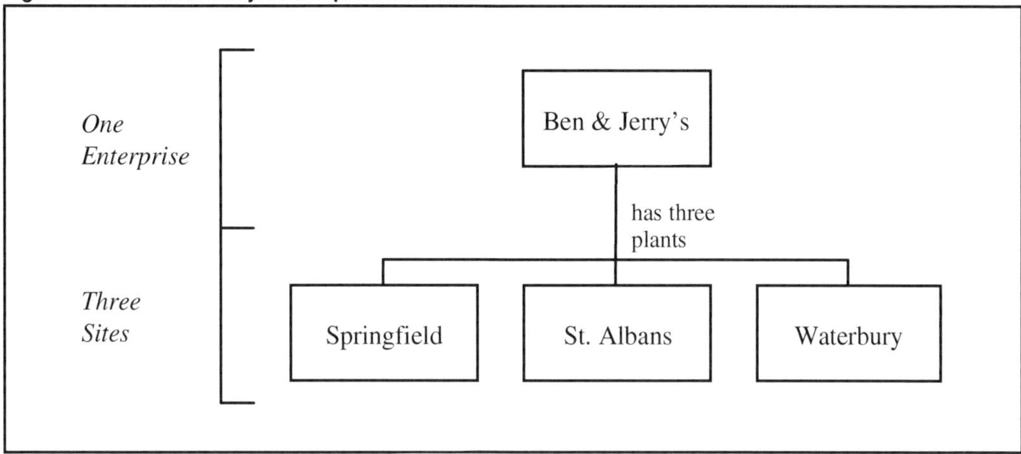

Regardless of where the ice cream was made, the product had to meet the same manufacturing and quality specifications.

Area Level

Areas are sections of a site. Like sites, areas can be organized by various means, including by physical layout or business function. Under S88, not every section of a plant need be part of an area, especially if that section has nothing to do with batch control. Figure 4.3 shows the areas in St. Albans. There are other sections of the plant that are not part of an S88 area, including packaging, dry receiving, our ammonia engine (compressor) room, the office area, and the cafeteria.

Figure 4.3 St. Albans's Areas

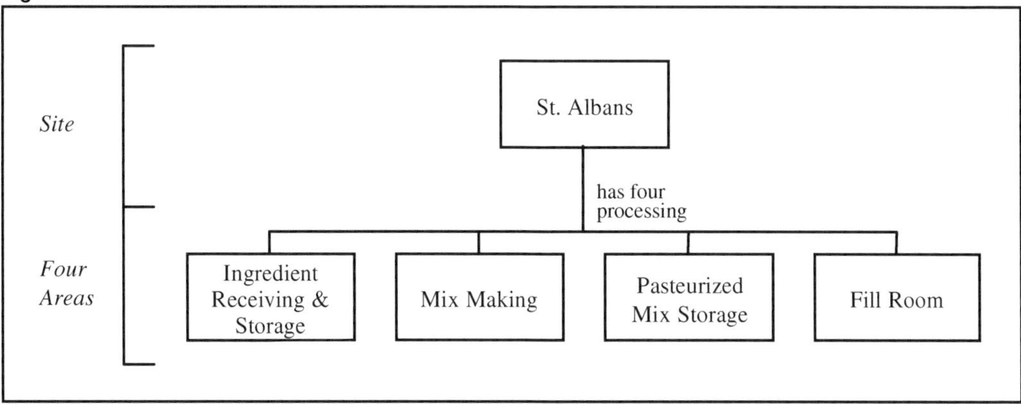

When you read the S88 standard, you'll find this wording in each of the sections describing the enterprise, site, and area levels: "There are many factors other than batch control that affect the boundaries of the [enterprise/sites/areas]. Therefore, the criteria for configuring the boundaries of a[n] [enterprise/site/area] are not covered in this standard." The S88 committee probably went to great pains in deciding how to organize and present this model, but the standard states that the top three levels are not considered beyond the initial definition. S88 is concerned with batch control, not business or political boundaries.

Pay attention now, we're skipping the *process cell* level and moving on to the *unit* level. We feel that we can explain the physical model better this way. We'll come back to the process cell level in a little bit.

Unit Level

Batching activities are focused on something called a *unit*. If you have faith in S88, batching cannot occur without the existence of units. The standard defines units this way: "One or more major process activities—such as react, crystallize, and make a solution—can be conducted in a unit. . . . [A unit] is usually centered on a major piece of processing equipment, such as a mixing tank or reactor." In other words, batch processing occurs *in* units. A unit combines ingredients, performs a reaction, or otherwise adds value to your product or interim product. The S88 definition mentions a mixing tank and reactor. Units are commonly vessels like these, but don't narrow your definition of a unit to just a vessel. A unit can be an in-line mixer, like a powder liquefier. (As we'll discuss later in this chapter, we chose not to make our powder liquefier a unit, but it could have been one.)

The term *unit* is somewhat abstract with respect to the equipment associated with it. For example, if you have a vessel that performs some major processing activity, you can think of the unit as follows:

- The vessel itself
- The vessel and attached instrumentation (such as level or temperature transmitters)
- The vessel and other associated equipment (such as valves, agitators, or recirculating pumps)
- Any combination of the above

Applying the S88 concept is much easier when equipment associated with a *single* unit is considered part of the unit. For example, if a reactor has a dedicated agitator (no other reactor uses that agitator), it is probably best to associate the agitator with the unit. Coordinating the control of the agitator is easier when it is permanently associated with the reactor. If it is not associated with the reactor, when the reactor needs it a request must be made to the batch management system to acquire the agitator as a resource. After the reactor is finished with the agitator, the reactor must release it. So, if a dedicated unit resource is not permanently associated with the unit, a lot of unneeded overhead is involved to use it. Inlet or outlet valves and instrumentation, such as temperature transmitters, are good candidates to be included as part of a unit. It is probably best to keep equipment not dedicated to a single unit separate or to include it as part of an equipment or control module. (We'll get to those in a minute.)

A good way to spot a unit is to determine if a piece of equipment must run a recipe in order to operate. If so, it is a unit. If not, it is used by a unit. Table 4.1 shows some examples of what we consider to be units, and what we don't:

Table 4-1. Unit Classification

Example	A Unit	Not a Unit
Mix-making batch tank	♦	
Pasteurizer	♦	
Reactor	♦	
Pump		♦
Ingredient storage tank		♦
Washing machine	♦	
Kitchen blender	♦	
Refrigerator		♦
Dishwasher	♦	

Even though we did not use OpenBatch to control the pasteurizer, we considered it to be a unit. Pasteurizers are typically three-section plate heat exchangers, and they can be large enough to hold hundreds of gallons of mix at any one time as part of a continuous process. Even so, we really didn't consider our pasteurizer to be a vessel. Whether or not you might think of a pasteurizer as a reactor, it certainly adds value to the product. After all, what real market value is there for unpasteurized ice cream? (Well, to pathologists, physicians, and lawyers there might be a lot of market value to unpasteurized ice cream.) Starting, running, and stopping a pasteurizer are no trivial tasks. Using a recipe to operate a pasteurizer is certainly valid, even though the "running" portion of the recipe is active much longer than the "starting" or "stopping" portions.

Pumps aren't units by themselves; they just move things around, and you typically don't need an entire recipe to operate a pump. (As we discussed earlier, a pump can be included as part of a unit.) We don't consider storage tanks to be units either. Remember that *a unit performs a major processing activity,* and the last time we checked storing didn't do that. Washing machines and kitchen blenders are units; we don't think refrigerators are. While some food, like pudding, may use a refrigerator to react, a fridge is really nothing more than a storage place. We think a dishwasher is a great example of a unit since it more or less runs a recipe, and we consider cleaning dishes and silverware as adding value. (If you disagree, that's fine; but please don't invite us over for dinner.) Here are some assumptions S88 makes about units, along with some comments from us that show how we interpreted those assumptions:

- *A unit frequently contains or operates on a complete batch of material*—In the St. Albans mix-making area, there were two units: *batch tanks* or *mix tanks*. Each batch tank was a receptacle for mixing multiple ingredients. Both batch tanks had outlet valves, multiple inlet valves, a mixer (agitator), and appropriate instrumentation.
- *A unit may contain or operate on only a portion of a batch*—If your batches are large, you may run several units in parallel, performing the same activities at once, or you may have two or more units in a *process train*, each sequentially working on a portion of a batch. At St. Albans, our two batch tanks operated in parallel, but we only made a batch in one of them at a time. This statement from the S88 standard also worked for us when we considered how we transferred mix to our single pasteurizer. From the moment the unit started transferring to the moment it was empty, the unit contained only a portion of the original batch.
- *A unit does not operate on more than one batch at a time*—From a record-keeping point of view, this made sense. It's hard to track batches or lots if you combine two or more of them into one unit.

Defining your units, as with many S88 issues, really depends on your interpretation of the standard. The three guidelines just outlined are based on what we think of units, but you may wish to expand your own definition. However, the number of units generally translates into the number of items that need to be communicated between your batch management system and your PLCs or DCS. (These communication "items" are often referred to as "tags.") Pricing for S88 software, like OpenBatch, is based on the number of tags needed. If you get unit-happy in your design, it'll probably cost you. A word of caution though: don't scrimp on units just to save some bucks on software. The nonmodular design that could result from your money-saving efforts could cost you much more in the long run.

Process Cell Level

A *process cell* contains all of the equipment, including units, required to make batches. Sometimes the term *train* is used to describe the units and all other equipment used to make a batch. A process cell may have more than one train, and the order of equipment used to make a particular batch is called a *path*. Here are some points to remember about trains:

- *A batch need not use all the equipment in a train*—Different products may require different sets of equipment, or different equipment may have different capacities. For example, one path through a train may require units 1 and 2 because they only hold 1,000 gallons each, while another path through the same train requires only unit 3, as it holds 2,000 gallons.
- *A train may be processing more than one batch at a time*—However, remember that S88 assumes that units only work on a single batch at a time.

- *A train cannot cross a process cell boundary*—A process cell is responsible for taking ingredients (raw or work-in-process [WIP]) and transforming them into finished products or WIP products.

Figure 4.4 shows the process cells in the St. Albans mix-making area.

Figure 4.4 St. Albans's Mix-Making Process Cells

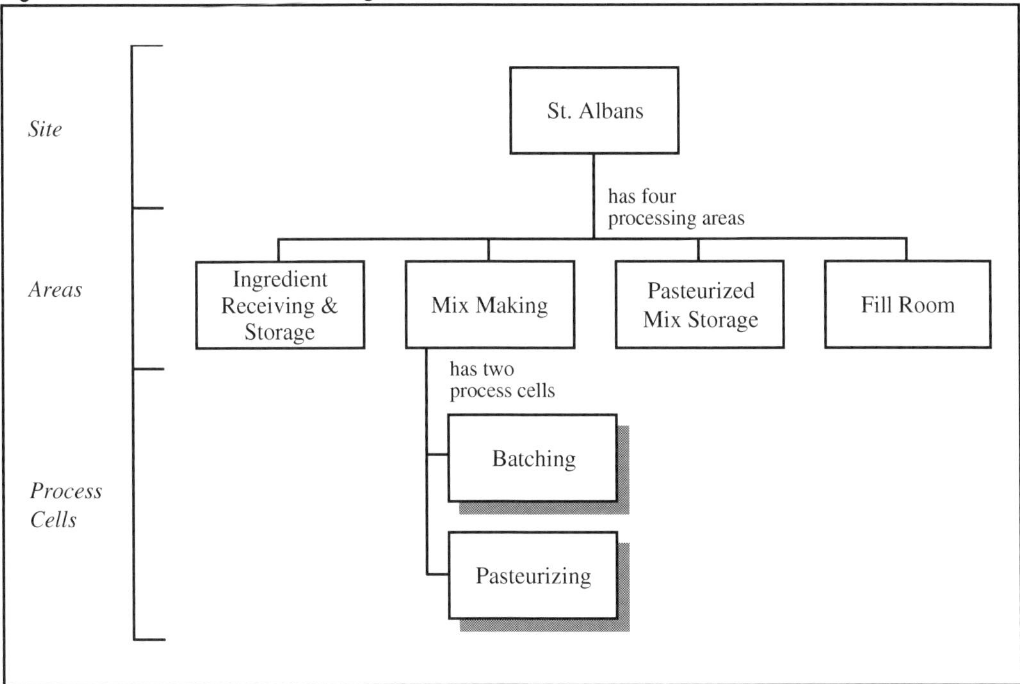

Deciding upon process cells really depends on how your process works. If you want to effectively manage the making of your batches, all of the resources needed to make batches should be within the process cell. Otherwise, a request to allocate resources will have to go outside the "domain" of the batch management system to some "foreign" system, complicating your operations. You can use your definitions of finished or work-in-process (WIP) products to help establish your process cell boundaries. Both the St. Albans units, the two batch tanks, fed a single pasteurizer (and related equipment such as the homogenizer), so it made sense to us to define our HTST pasteurizing process as a separate cell.

Trains within process cells can have three different structures: single-path, multiple-path, and network-path. Figures 4.5, 4.6, and 4.7, respectively, illustrate these. Note that our mix-making system in St. Albans used a multiple-path structure.

Figure 4.5 Single-Path Process Cell Structure

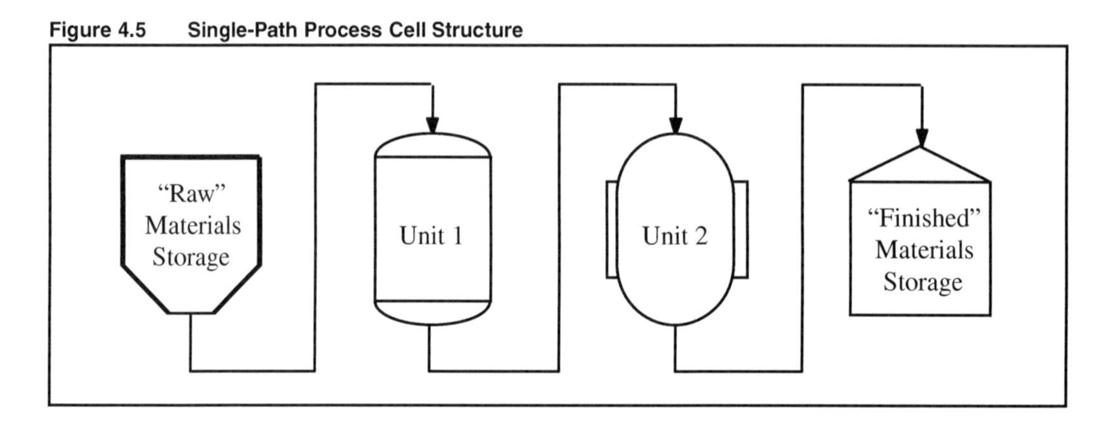

Figure 4.6 Multiple-Path Process Cell Structure

"Raw" Materials Storage

Unit 1

Unit 2

"Finished" Materials Storage

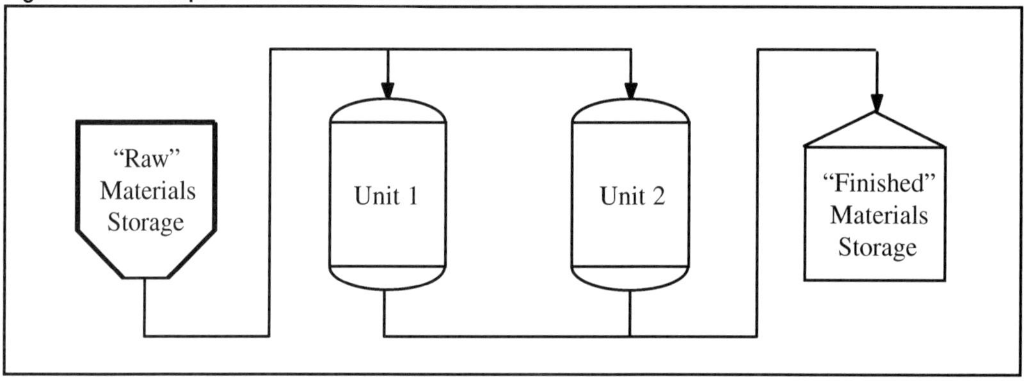

Control Module Level

Time to pay attention, again. We just skipped the *equipment module* level to help explain the physical model. We'll get to equipment modules in the next section. *Control modules* are the most basic element of the physical model. The S88 standard defines a control module as "typically a collection of sensors, actuators, other control modules, and associated process equipment that, from the point of view of control, is operated as a single entity." So, from a control standpoint a control module is treated as a *single entity*. Each control module provides a direct "connection" to the process through actuators and sensors. For example, let's say you want to control the rate of flow of some liquid. You use a pump driven by a variable-speed motor and a flowmeter. A proportional-integral-derivative (PID) controller reads the flow rate and sets the pump speed appropriately. The batch control system wants only to control the flow rate and considers the combination pump, flowmeter, and PID loop as a single control module. Keep in mind that even though the control module exists in the physical model, not all elements need be physical. In our example, the PID controller can be a PLC instruction or DCS object and not a stand-alone device that physically links the flowmeter to the pump.

Figure 4.7 Network-Path Process Cell Structure

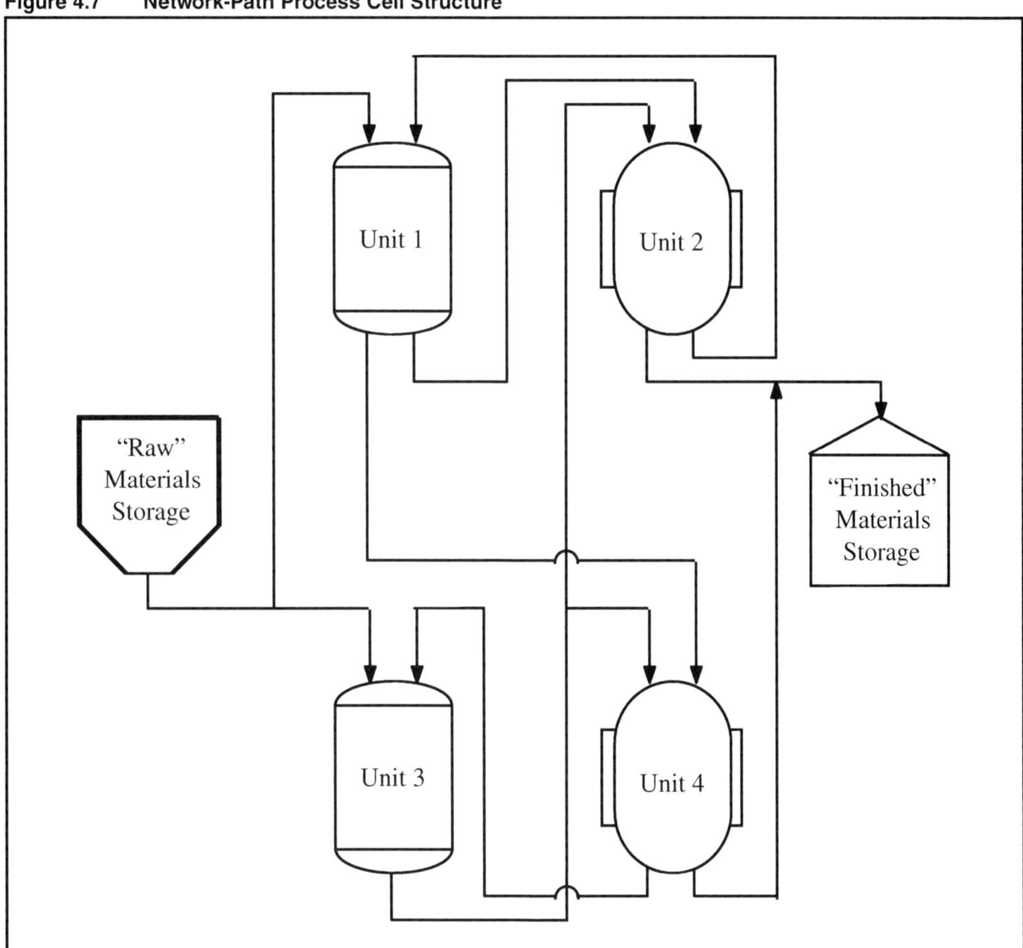

Another example of a control module may be a group of valves on a header that direct flow to one of many destinations based on input to some control function. Again, the control function may exist in a DCS or PLC. The same control module can be used by different functions, such as charging different ingredients, or in the production or clean-in-place (CIP) modes.

In the simplest form, control modules can just be device drivers, but they can do so much more for you. Control modules should provide a robust method of device control, including these functions:

- *Automatic and manual modes*—Automatic mode means PLC or DCS logic or some stand-alone function controls the device. Manual mode implies that the device is controlled directly by an operator.
- *A simulation mode*—This allows an operator or engineer to test software without actually manipulating equipment.

- *Permissives*—These prevent a device from actuating if constraints or permissions are not true. For example, a tank outlet valve cannot open if the adjoining header is being cleaned in place.
- *Alarms*—These provide feedback on a device's operation. For example, an alarm is set if a valve is commanded open, but a limit switch does not confirm that it has opened.

For example, PLC or DCS logic may command a valve to open via a control module, but if a simulation bit is set the valve will be prevented from opening from within that same control module. Limit switches attached to the valve may provide feedback to the control module for "hard" alarms, while the use of permissives allows for "soft" alarms. Figure 4.8 shows one possible design for a control module.

We consider two issues to be especially important with regard to control modules. First, instrumentation can serve more than one control module. For example, one flowmeter can be used by a control module that charges water into a batch tank and another control module that charges cream. Second, a piece of equipment is controlled by one (and only one) control module. There may be several logical events that trigger the control module to manipulate the equipment. For example, a pump can be started by production logic or by CIP logic. The Auto Command Enable bits in Figure 4.8 show this. (Some control engineers may wish to implement a control module by latching or unlatching the Auto Command bit directly, without using the enable bits.)

Figure 4.8 A Sample Control Module

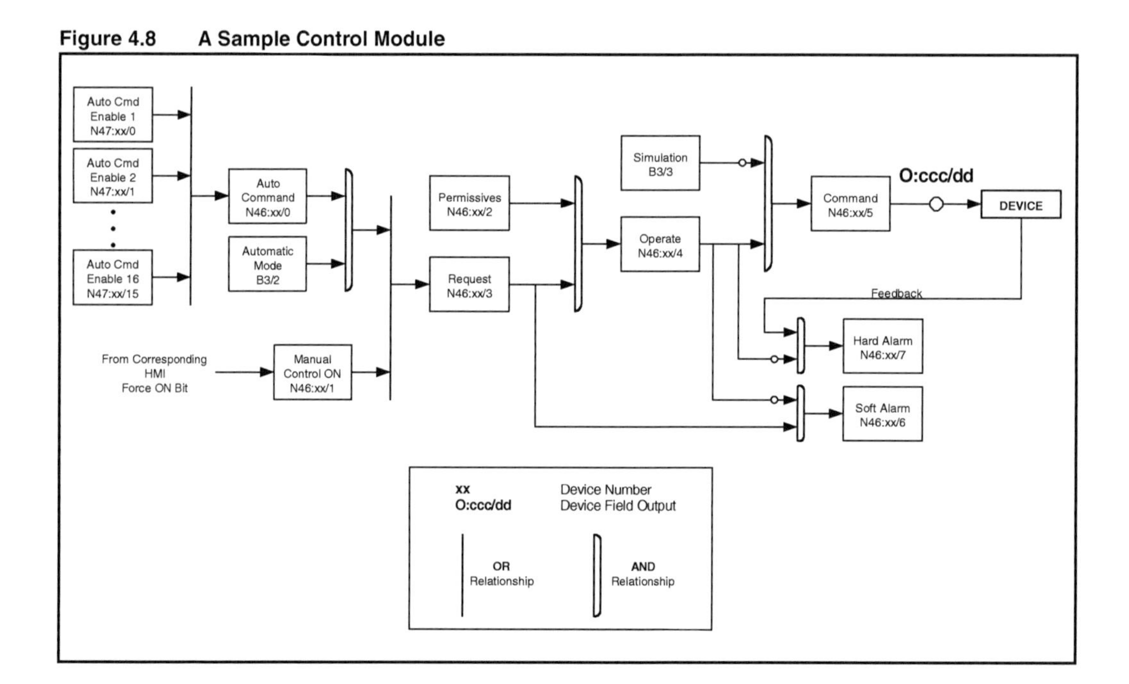

Equipment Module Level

The S88.01 standard defines an *equipment module* as "a functional group of equipment that can carry out a finite number of specific minor processing activities." In other words, equipment modules *group* physical devices for performing one or more specific functions. An equipment module may be made up of control modules or other equipment modules.

S88 implies that equipment modules can include decision-based logic. We agree, but at Ben & Jerry's we kept it very simple. For example, St. Albans incorporated cocoa in batches through a powder liquefier that mixed the cocoa with a liquid ingredient passing through it. If the liquid flow rate fell below a certain limit, the cocoa might plug the powder liquefier. If this happened, the operator had to hold or stop the batch and physically clean the liquefier, a time-consuming and costly process. Therefore, the PLC logic had to constantly check for the proper flow rate and close the cocoa valve if the rate was too low.

An equipment module can also work with other forms of instrumentation. Plants often have several storage tanks for the same ingredient. Sometimes, each tank has its own transfer line, pump, and flowmeters. Because each holds the same ingredient, you can choose to write a single set of PLC logic instructions and use an equipment module to control the correct devices. Not only can you control valves and monitor low-level switches as described earlier, but you can also copy the different flowmeter totalizer values into a common register based on the raw material tank being used to supply the ingredient. This way, the PLC logic need only look at the common register to see if the targeted amount of sweetener has been transferred.

Different units can share equipment modules. This makes sense if you have ingredient storage tanks that share a common transfer line. If an equipment module can be active only with one unit at a time, it is called an *exclusive-use resource*. If it can be active with several units simultaneously, it is called a *shared-use resource*. Certainly, the converse is true also: there can be equipment modules used exclusively by a single unit.

Equipment modules can be permanently "included" as part of a larger grouping, such as a unit or another equipment module. They can be temporarily "attached" as part of a larger grouping, or they can stand alone. If operated stand-alone, an equipment module may be used by only one unit at a time (exclusive use) or by many units simultaneously (shared use). The same is true of control modules. They can be permanently or temporarily part of other control modules, equipment modules, or units. Any temporary attachments are formed when a batch is run. For instance, the process cell might have an ingredient recirculation loop that is common to both batch tanks. The recirc loop can be active while a recipe is running, or the operators can run it "manually" as a stand-alone function. Since the plant only wants to "recirc" the contents of one unit at a time (the operators wouldn't want to mix the contents of two batch tanks), the recirc loop equipment module is an exclusive-use resource.

You may be asking a pretty big question right now: "Why couldn't the logic of equipment modules be placed in control modules"? That's a great question, and in some cases control modules can replace equipment modules. Control modules are subordinate to equipment modules only by virtue of the fact that equipment modules trigger actions that cause control modules to physically manipulate equipment. However, if equipment modules aren't needed, why add another level of complexity? There is one big difference between control modules and equipment modules, and it has to do with the type of software logic S88 says they're allowed to run. If you can wait, we're going to answer these questions in Chapter 5.

Designing the Physical Model

Designing the physical model isn't that hard. If you are designing a "green field" site or adding new area or process cell to an existing site, you have the luxury of purchasing and locating equipment according to your S88 plans. If you are applying S88 to an existing process, as we did, you may have to think a little differently. Fortunately for us, applying S88 wasn't too difficult. We're assuming that it won't be much different for others retrofitting S88. This is due to the robustness of the standard and the foresight of the committee members.

We're going to start from the top. Let's consider the enterprise level as a given. It's your company, business unit, or division. The site level probably is fairly easy also. Just work from a geographical or four-wall approach when defining your sites.

If a site essentially manufactures one product, or one type of product, determining areas within it probably isn't too difficult. Look at your plant layouts and your process and instrumentation drawings (P&IDs) to understand issues like material flow. Figure 4.3 shows the areas in our plant that are associated with process manufacturing. If a site manufactures several types of products, plant layouts and material flows can still define boundaries. However, you may have to look at the operation from a business perspective and see what equipment or site functions each product or business unit "owns."

Process cells take shape when you further break down the process required to make your products. Consider how you run your site business and keep looking at your P&IDs to help you determine your process cell boundaries. Remember that batches utilize one or more units in a train. A process cell can contain more than one train, but according to S88 a train is not allowed to cross process cell boundaries. *Use this fact as an easy test to see if you've defined your process cells too narrowly.* S88-aware software, following the standard, starts considering the physical model at the process cell level. Some S88 software packages refer to an "Area Model" because it is the level that encompasses process cells.

We chose to place our batching process and our pasteurization process into two separate cells. The link between these cells was a single transfer pipe. Even though we had two units in our batching cell, mix could be made in one or the other, but not both. Because our batch tanks then operated in parallel (though not simultaneously), they were not considered part of a train.

P&IDs are probably your best source of information for determining units. Consider how you run your process. Units are equipment that perform major process activities and often work on an entire batch at a time. If you have a vessel that combines two or more ingredients, it is probably a good candidate to be a unit. Vessels that transform a product without necessarily adding more ingredients, such as a reactor, are also generally considered to be a unit. We don't believe storage tanks should be units because they add no value to the end product and often simply hold raw ingredients or WIP products. In our application of S88 at Ben & Jerry's, we also did not need a recipe to operate a storage tank.

As we've mentioned before, our plant only had two units in the mix-making area: the two batch tanks. During the design phase of our project, we seriously considered whether to make our powder liquefier a unit. After all, the powder liquefier combined several ingredients and performed a major processing activity. Our decision to designate it as a piece of equipment and not a unit was based on several things:

- The powder liquefier can be active when any two of five liquids are transferring. Not all liquids are present in every recipe.
- Depending on the recipe step, one of those five liquids must sometimes bypass the liquefier and transfer directly to a batch tank. (As the liquefier produces significant shear, we didn't want cream passing through it.) We chose to have cream and milk transfers be handled by the same PLC logic because both ingredients come from the same storage tank area and pass through a common cluster and header. If the powder liquefier were a unit, we would have needed two different sets of PLC logic to run nearly identical equipment.
- To keep our recipes as simple as possible, we wanted to maintain only a single unit per batch. As you'll read in Chapter 5, recipes become more complex hierarchically as you add units.

Here's an important point: the physical model is collapsible. For example, a control module need not be part of an equipment module for it to be associated with a unit. The S88 physical model as shown in Figure 4.1 only has one "must contain" label. Process cells must contain at least one unit; all other associations are optional. But if you truly want to be part of the S88 club, all control devices must be part of some physical entity: a process cell, a unit, an equipment module, or a control module.

Keep one important thing in mind when designing your physical model. You should create your model based on equipment, not ingredients. For example, cream and milk ingredients can be stored in one of several general-purpose raw dairy storage tanks. (You may not want to impose additional constraints on your process by limiting what ingredient can go into a particular tank.) We decided to create a single set of PLC logic instructions and pass an ingredient parameter to it because all the equipment used to transfer the ingredients was identical for both ingredients.

As you've just learned, the physical model describes the equipment necessary to make a batch. However, batch processing requires that equipment actions be sequenced. Recipe procedures handle this sequencing, and that is what Chapter 5 is all about.

Recipes, Part 1: Procedures

If we could have stuck a couple of acetaminophen tablets on this page, we would have. If any chapter in this book is likely to create headaches, we think it's this one. You may want to grab a couple of tablets just in case. (Acetaminophen is made using a batch process you know.)

As you read in Chapter 4, the physical model describes the equipment necessary to make a batch. However, we need more than equipment to make a batch. Batch processing requires that equipment actions occur in a defined sequence. That, more or less, is the primary function of a *recipe*. This chapter and the next are all about recipes. In this chapter, we're going to focus on recipe procedures. In Chapter 6, we will discuss all other aspects of recipes.

According to S88.01, a recipe is defined as "the necessary set of information that uniquely identifies the production requirements for a specific product." We've all used recipes before to make scrumptious things like marinades and desserts. Of course, not all products made from recipes are edible. The last time we checked, shampoo didn't taste very good. Eating it isn't very nutritious either. From a different perspective, if you've had projects in your career like some of ours, you've also had recipes for disaster.

Information in a Recipe

According to the founding fathers (and mothers) of S88.01, a recipe contains five categories of information. They are shown in Table 5.1.

Table 5-1. Recipe Contents

Header	Administrative information and a process summary
Equipment Requirements	Information about the specific equipment necessary to make a batch or a specific part of the batch
Procedure	Defines the strategy for carrying out a process
Formula	Describes recipe process inputs, process parameters, and process outputs
Other Information	Product safety, regulatory, and other information that doesn't fit in the other categories.

This chapter is all about procedures. We will discuss the other four categories in Chapter 6. As listed in Table 5.1, according to S88 a recipe *procedure* defines the strategy for carrying out a process. That is, the procedure explains how ingredients (raw materials) are to be combined, reacted, or otherwise processed to make a batch. The procedure describes a structured sequence of processing activities necessary to make a product. Don't think about equipment in this chapter; think about running a functional procedure.

Types of Recipes

Different parts of an enterprise (e.g., your company) may require different types of information about a product or the process to manufacture it. Consider this scenario:

- An R&D group is probably concerned with the properties of the product and general processing procedures necessary to make it but may not care exactly what specific equipment is used.
- The corporate engineering group tasked with creating a process to manufacture the product may be very focused on the type of equipment used but not the specific piece of equipment used per batch.
- A site production team is very interested in what specific pieces of equipment are available at a certain time to make a batch.

Each of these three groups needs different types and amounts of information. Using only one recipe to hold the necessary information for all three groups would be complex and cumbersome for everyone, so S88 defines four types of recipes that focus on varying levels of specific information. Figure 5.1 shows the four recipe types.

The *general recipe* is used at the enterprise level and is the basis for lower-level recipes. It defines raw materials and their quantities as well as the required processing to make the product. However, the general recipe is created without specific knowledge of a particular site or the equipment that will manufacture a product; it is supposed to communicate processing requirements to multiple manufacturing locations. For example, since the three Ben & Jerry's plants were built years apart, the equipment in each plant sometimes differed. However, the basic process function the equipment had to perform, such as incorporating cocoa in a batch, was identical. The general recipe is only concerned with process functions, not equipment. A general recipe in this case may state *Blend cocoa into liquid ingredient LI2080*, not *Use blender BL22 to incorporate cocoa into liquid ingredient LI2080*. In some cases, process requirements may require specific equipment, so the general recipe may have to call out equipment. People with knowledge of both the product characteristics and processing requirements—such as an R&D group—create this recipe.

Figure 5.1 S88 Recipe Types

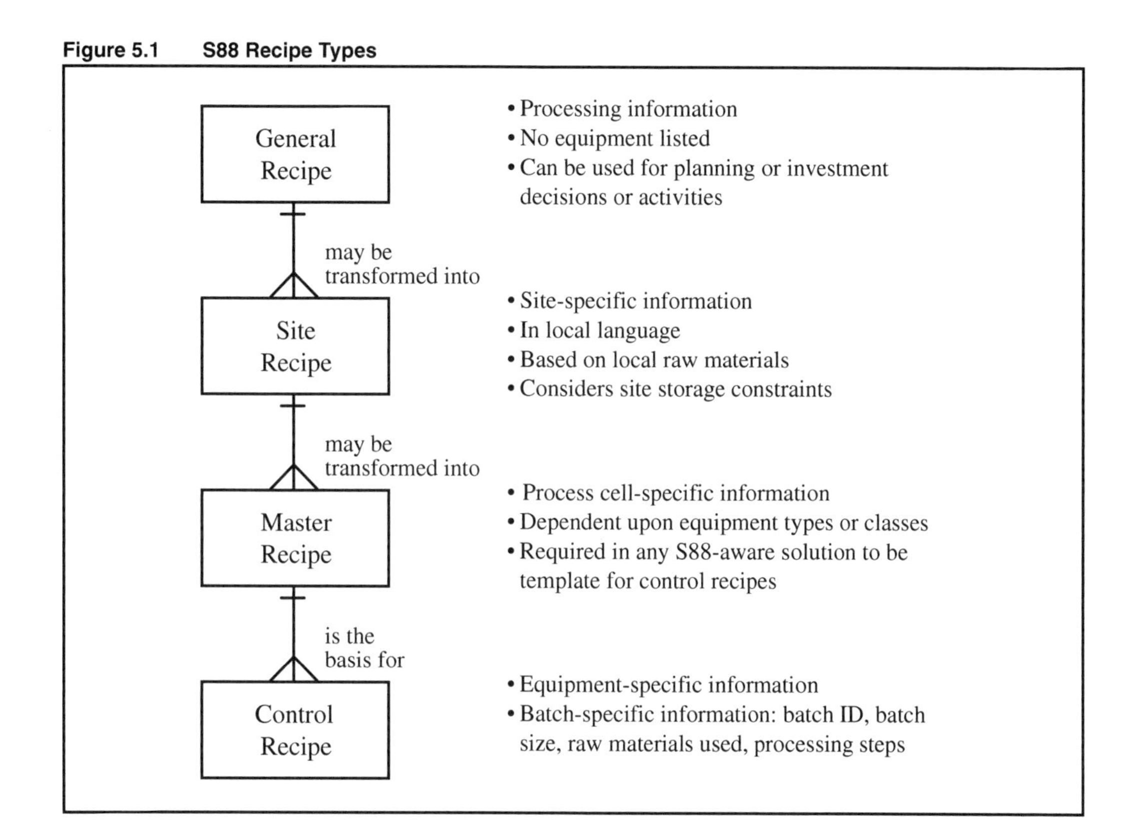

The quantities expressed in the general recipe may be fixed, or they can be normalized to a standard size to be scaled up at plants. Our mix recipes were all normalized at somewhere around 100 gallons. However, our plants made batches much larger than that, so recipes were scaled up during production. Equipment requirements are expressed in terms of attributes needed, such as materials of construction (e.g., a stainless steel versus a galvanized steel vessel) or processing characteristics (e.g., maximum mixing shear or pressure range).

The general recipe can also be used as a basis for company planning or for investment activities. The information contained there may be used to plan production or to provide information to customers or regulatory authorities. Food companies may use the general recipe to form the nutrition information and ingredient decks that are required on product packaging. Chemical companies and even food ingredient suppliers can use information from a general recipe to create a material safety data sheet (MSDS).

As its name suggests, a *site recipe* is specific to a manufacturing site. It is usually derived from the general recipe to meet the specific conditions or constraints of the site manufacturing the product. The site recipe provides the level of detail necessary for long-term production scheduling for a site. Perhaps a corporate engineering group produces site recipes from a general recipe. S88 does not require the existence of a general recipe, however, so a site recipe can be created without a corresponding general recipe.

Multiple site recipes are often created for the convenience of plants that are separated by large geographic distances. Site recipes might differ as to the language in which they are written or the specifications for local raw materials. A site recipe for a plant in Germany, for example, should probably be written in German and designed to accommodate raw materials from Germany or Europe. While it may still not be specific to a particular process cell or set of equipment, a site recipe may be specific to on-site processing or storage capacity and constraints.

A *master recipe* is targeted to a process cell or a subset of the equipment in a process cell and is derived from either a general or a site recipe. Master recipes depend more on equipment types or classes, such as a temperature-controlled vessel or a medium-shear mixer. Differences in instrumentation can be called out in different master recipes. For example, one process cell may dispense an ingredient using load cells, while another may use flowmeters. A plant control engineer (with or without the help of corporate engineering) may transform a site recipe into one or more master recipes.

Like the general and site recipes, the master recipe's quantities may be specified as normalized, fixed, or calculated values. It can contain product-specific information required for detailed scheduling, such as equipment requirements. But unlike the general and site recipes, S88 batch control *requires* a master recipe. A master recipe is the template for recipes used to create individual batches. Without this template, no specific batch recipes can be created, and therefore no batches can be produced.

A *control recipe* is used to create a single, specific batch. It starts as a copy of a specific master recipe and is modified as necessary to create a batch. The modifications may account for batch size, the characteristics of raw materials on site, or the actual equipment to be used. While several (or dozens, hundreds, or thousands of) batches may use the same master recipe, every batch has a single control recipe unique to that batch and that batch alone. Two control recipes may be identical in terms of ingredients, quantities, or equipment used, but they are identified individually nonetheless. Control recipes unique to individual batches allow product tracking or genealogy to occur.

For example, ice cream companies may have master recipes that assume the potency of the fat and nonfat solids in their cream and milk. However, since some dairy companies do not order dairy products standardized to specific potencies, each raw dairy storage tank most likely will have different potency values. This means that control recipe quantities will vary from the master recipe if the raw dairy tank potencies are different from those the master recipe specifies (as often happens). Also, the master recipe does not care from what raw dairy tank cream or milk is transferred. However, for reasons related to product tracking and process variability, the raw dairy tank that is used needs to be recorded. This is captured in the control recipe. Companies also capture other important information, such as which batch tank is holding the mix and the desired speed of the tank agitator.

Figure 5.2 shows an exploded view of a recipe managed in a three-site enterprise. Note that plant 3 only has one process cell. In this case, the formal need for a site recipe may no longer exist, but the recipe might be maintained for consistency with all the sites.

Figure 5.2 Exploded View of a Recipe

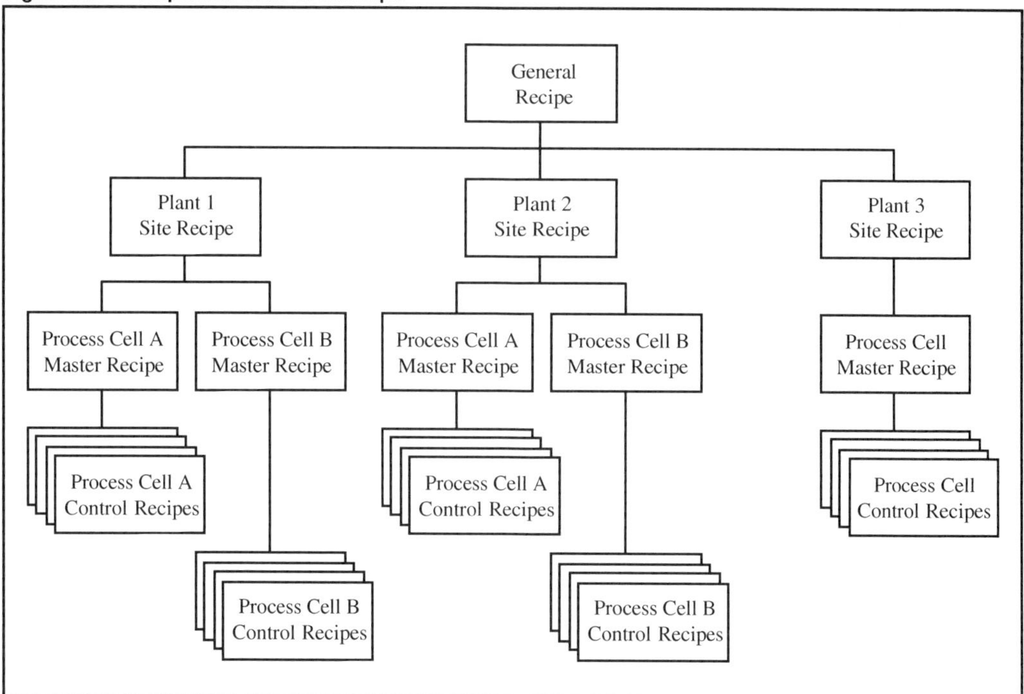

Many S88-aware batch control solutions, including OpenBatch, start managing recipes at the master recipe level. Engineers or lead operators may define master recipes. When a batch is scheduled for production, OpenBatch creates a control recipe in memory and downloads batch-specific information to a DCS or one or more PLCs. OpenBatch creates a batch record for each control recipe, storing time-stamped default and user-specified information about the batch as it is made. As an option, it can also archive the control recipe batch record to any ODBC-compliant database. (*ODBC* stands for "Open Database Connectivity.") When the batch is finished and the operator removes the batch from the active batch list, the control recipe no longer exists, but the data collected from the recipe and batch remains.

General and Site Recipe Procedures

Two different models are used to define recipe procedures, depending on the recipe level. (Ha! We're sneaking in two more models.) Since the general and site recipes focus on processing activities rather than referring to specific equipment, the procedures for the top two recipe levels are based on the *process model*, as shown in Figure 5.3.

Figure 5.3 General and Site Recipe Procedures Are Based on the Process Model

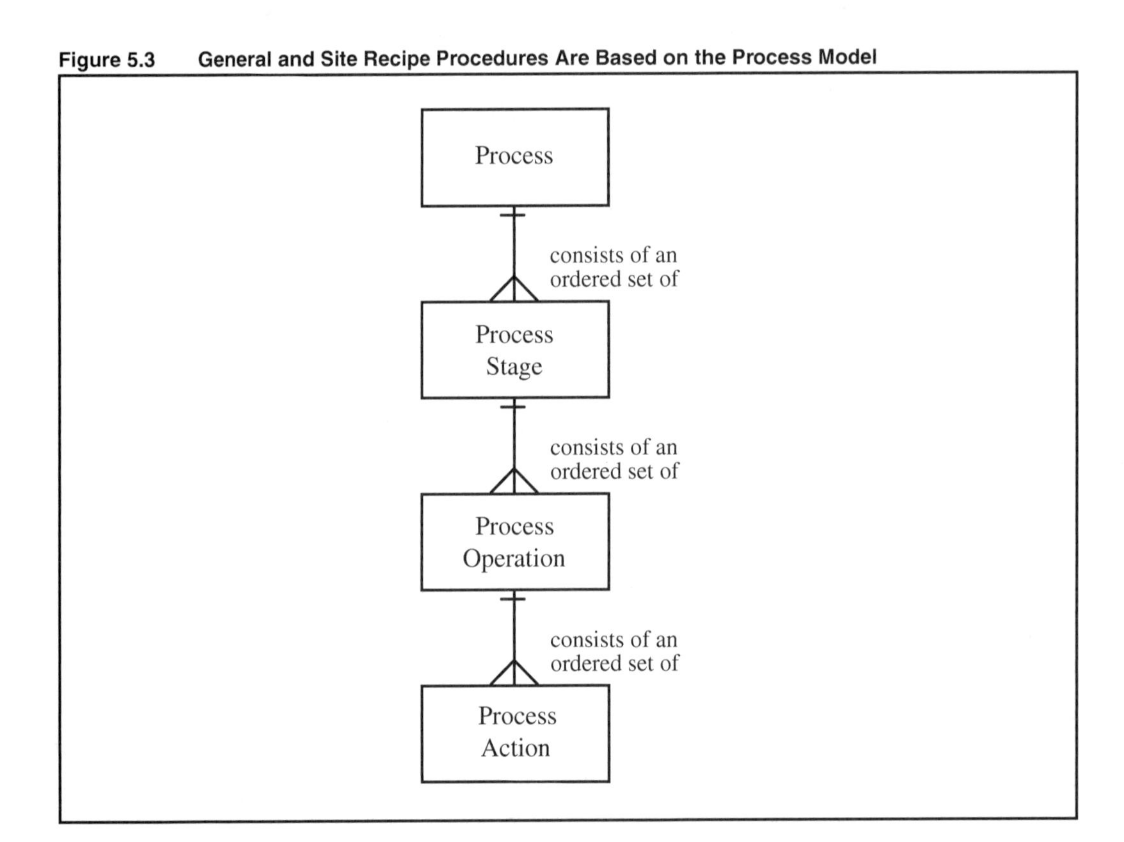

The S88 process model is very conceptual and describes the functionality that is necessary to create a batch. When we describe batching using the process model, we really don't care about equipment or detailed procedures. The model may be a tool used by your product R&D group, which needs to describe the steps that must occur in order to make a product.

The top level of the process model represents the overall process functionality needed to make a batch. Recall from Chapter 1 that we defined a batch process as one that "leads to the production of finite quantities of material by subjecting quantities of input materials to an ordered set of processing activities over a finite period of time using one or more pieces of equipment." The process model describes these process activities.

Each batch process is made up of one or more *process stages*. Process stages can run one at a time or in parallel, but they usually operate independently of one another. If you're into chemicals, Table 5.2 illustrates typical process stages for making polyvinyl chloride (PVC):

Table 5-2. Process Stages for PVC

Process Stage	Description
Polymerize	Polymerize vinyl chloride monomer.
Recover	Recover residual vinyl chloride monomer that did not polymerize.
Dry	Dry the polyvinyl chloride.

One or more *process operations* make up a process stage. A process operation represents a major processing activity and generally results in a chemical, physical, or biological change in the material being processed. Typical process operations for the *polymerize* process stage may include those shown in Table 5.3.

Table 5-3. Operation of a Process Stage

Process Operation	Description
Prepare Reactor	Evacuate the reactor to remove oxygen.
Charge	Add demineralized water and surfactants.
React	Add vinyl chloride monomer and catalyst, heat 60°C, and hold until reactor pressure decreases.

Each process operation can be subdivided into *process actions*. Process actions describe minor, independent processing activities that serve as building blocks of the process model. The *React* process operation in making PVC might have the process actions shown in Table 5.4.

Table 5-4. Actions in a Process Stage

Process Action	Description
Add	Add required amount of vinyl chloride monomer to reactor.
Add	Add required amount of catalyst to reactor.
Heat	Heat reactor contents to 60°C.
Hold	Hold reactor contents at 60°C until reactor pressure decreases.

In making mix at Ben & Jerry's, we used the process actions shown in Table 5.5 as part of the process operation that creates a batch.

Table 5-5. Actions in the Mix-making Process Operation

Process Action	Description
Add Water	Add required amount of water to batch tank.
Add Sweetener	Add required amount of liquid sugar to batch tank.
Add Milk	Add required amount of milk to batch tank.
Add Cream	Add required amount of cream to batch tank.
Add Eggs	Add required amount of liquid egg yolks to batch tank.
Agitate	Blend ingredients.

Because the process model is not based on equipment, process stages, process operations, and process actions are not constrained by unit boundaries. For example, a single process stage may execute in several units. The procedure information in a site recipe generally has a one-to-one relationship with the procedure information in the corresponding general recipe. That is, a process stage in a general recipe is equivalent to the process stages in all of its derived site recipes, and so on. As with other site recipe information, however, the process stages, process operations, and process actions may be modified to meet site-specific requirements.

Master and Control Recipe Procedures

When considering recipe procedures, a transformation occurs between the site and master recipe levels. Unlike general or site recipes, master and control recipes deal with equipment classes or specific pieces of equipment. Therefore, the process model-based site recipe procedure must be converted to a recipe procedure that is based on the *procedural control model* in the master recipe. The model in Figure 5.4 shows this master and control recipe procedure, which is specific to a process cell.

A *procedure* is the highest level in the procedural control hierarchy. It defines the overall strategy for making a batch. It consists of an ordered set of unit operations and must exist if more than one unit is used to complete a finished batch.

A *unit procedure* is an ordered set of operations that is carried to completion on a single unit. That is, a unit procedure is a contiguous production sequence acting on one and only one unit. Material cannot be processed in one unit, transferred out to another unit, processed there, and then returned to the first unit all within the same unit procedure. Only one unit procedure is allowed to be active on a unit at any time. Multiple unit procedures can run concurrently as part of the same procedure, as long as they are active on different units.

Figure 5.4 Master and Control Recipe Procedures Are Based on the Procedural Control Model

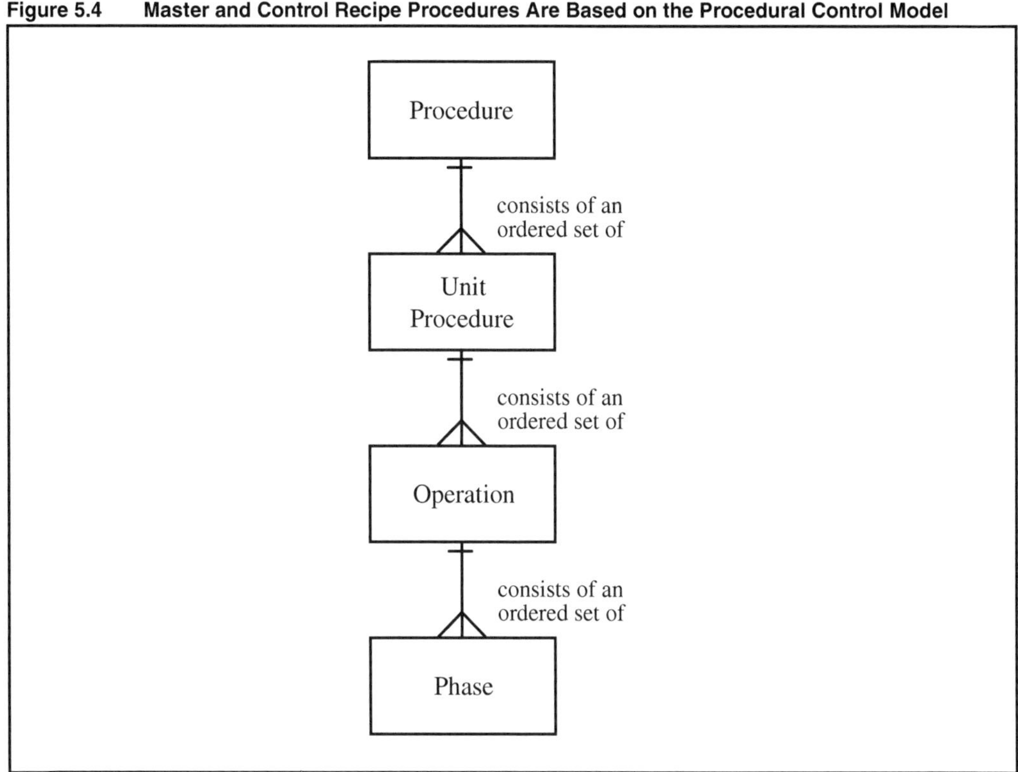

Figure 5.5 shows an example of two unit procedures running on two different units at once. A third unit procedure is run only after the first two complete. For fun, let's call it toothpaste, assuming that units 1 and 2 are both batch vessels and that unit 3 is some type of blender/extruder. Just in case you're wondering, we can assume that each unit procedure contains instructions that will handle the transferring of both types of toothpaste from the vessels to the blender/extruder.

Figure 5.5 Make Cool Swirled Toothpaste Procedure

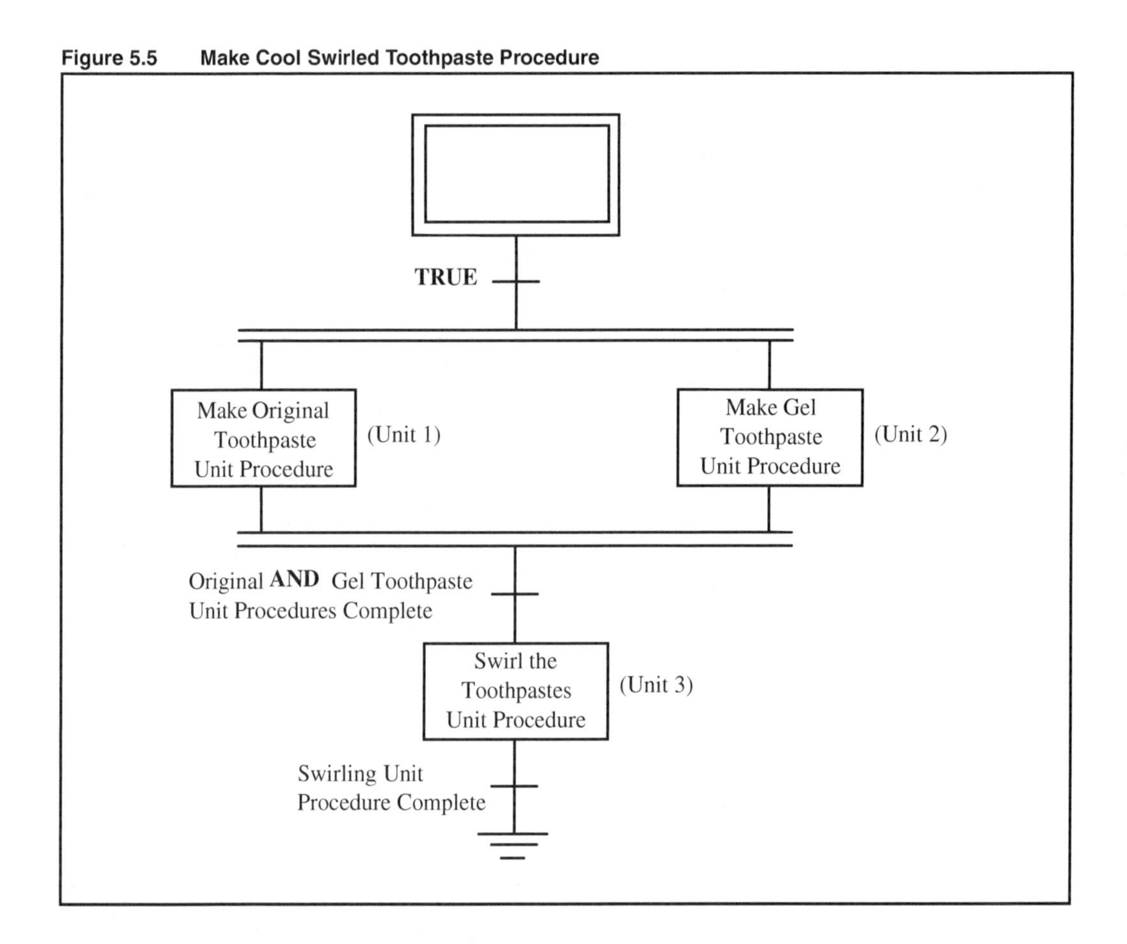

An *operation* is an ordered set of *phases* (which are described below) that are carried to completion within a single unit. Operations usually involve taking the material being processed through some type of physical, chemical, or biological change. If your process involves volatile reactions, keep in mind that S88 committee members believed that it is desirable to locate operation boundaries at points in the procedure where normal processing can be safely suspended. That is, design your operations so that when one completes a chemical or biological reaction is stabilized. Like unit procedures, the standard assumes that only one operation is active on a particular unit at a time. Figure 5.6 shows the detail of the *Make Gel Toothpaste* unit procedure from Figure 5.5.

The *Add Ingredients* operation combines various ingredients to form a new interim material. This operation processes the material through a physical change. The *React Toothpaste* operation mixes, heats, and then cools the new mixture to create the toothpaste. This operation processes the material through both a physical and chemical change. The *Prepare to Transfer* operation gets the vessel unit ready to transfer its contents to the blender/extruder unit.

Figure 5.6 Make Gel Toothpaste Unit Procedure

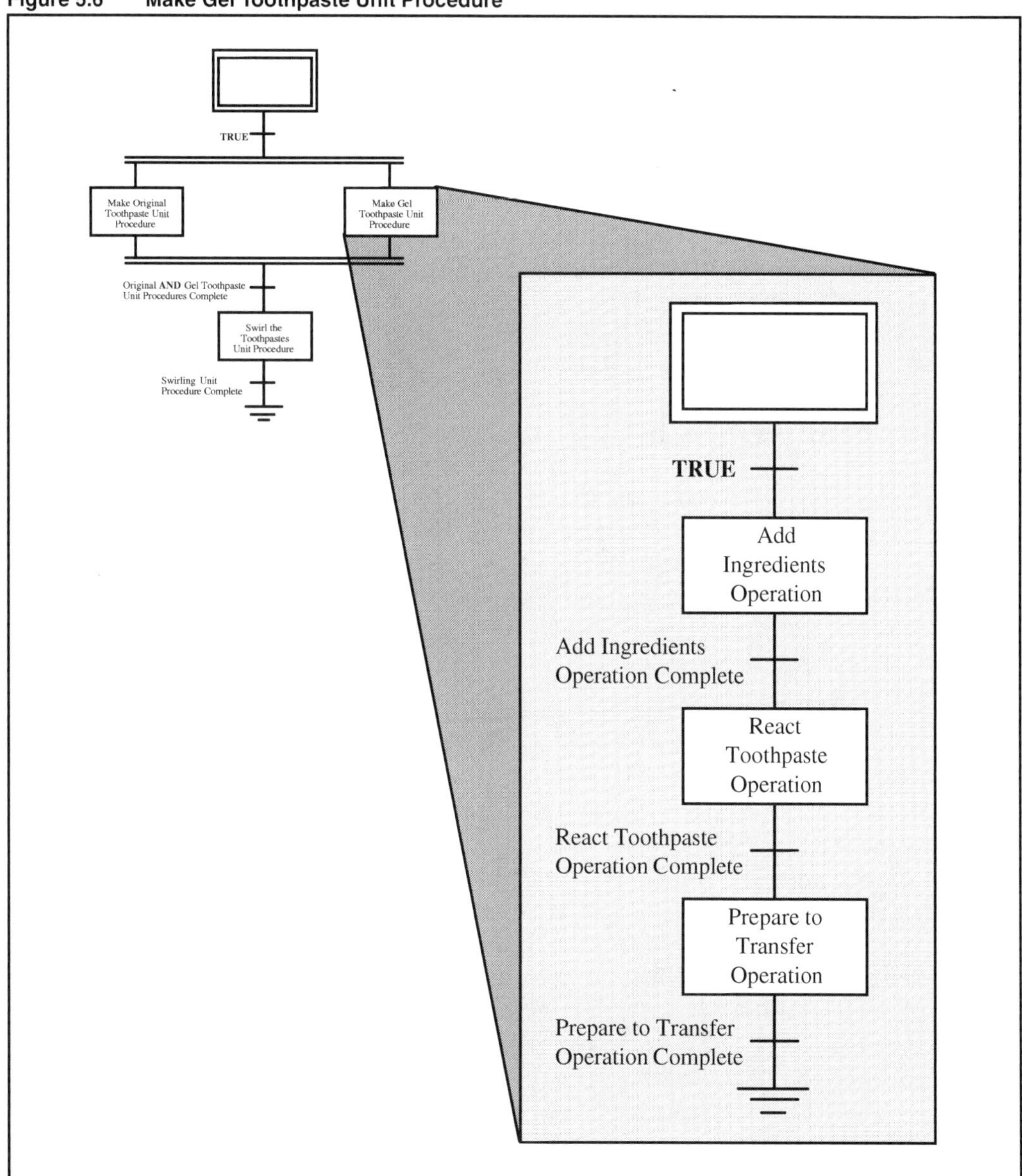

A *phase* is the smallest element of procedural control that can accomplish process-oriented tasks. Phases perform unique, basic, and generally independent process-oriented functions. If a phase controls more than one basic process-oriented function, consider splitting it into several phases and wrapping those into an operation. In making mix at Ben & Jerry's, our phases performed basic and independent actions, such as transferring sweetener and agitating a batch tank.

While we'll discuss this in more detail later, keep in mind that somehow the procedure to make a batch must link to physical equipment control. Often this is done at the phase level. In that case, people will refer to a *recipe phase* as an abstract term of control for use in recipes and will refer to *equipment phase logic* as the actual implementation of the phase. In real terms, phase logic is software that controls equipment.

Phase logic generally exists in process control equipment, such as a PLC or a DCS. A recent trend has been for phase software to also exist in PCs using PLC emulators or Visual Basic (VB). Phases can connect to equipment and instrumentation via traditional hardwired I/O or by using device-level networks, like DeviceNet, Fieldbus, Profibus, or Interbus-S. Since Ben & Jerry's had an existing Allen-Bradley PLC-5, all our phases existed as PLC ladder logic. Some day, the St. Albans plant may implement VB phases.

Here are some more rules about using phases:

- *A phase can operate on more than one set of equipment, such as different units, just not at the same time*—For example, OpenBatch allows users to "share" phases between units to eliminate unnecessary code duplication. This really helped us at St. Albans since our two batch tanks operated in parallel. Except for a few specific valves on each unit, all of the equipment associated with transferring product to the units was identical.
- *Physical entities, such as units, equipment modules, and control modules, can be acted upon by more than one phase*—For example, you may have a production phase and a CIP (clean-in-place) phase that both command the same pump control module. The control does not have to be mutually exclusive, however. In our batching system, several ingredients met in a common header before entering a batch tank. This header has a single valve at one end. Our raw dairy, water, and sweetener phases all opened this valve via a control module. If at least one of these phases was running, the valve was open. If more than one of these phases was running, each phase attempted to open the valve, and that was okay. If the raw dairy, water, or sweetener phase completed, it attempted to close the valve. If another one of these phases was still running, the valve wouldn't close, and that was okay too. Since our mix-making process required the valve to stay open in the event of a conflict, we designed the control module so that one phase attempting to close the valve would not override another phase that needed it open.
- *A phase may require that one or more other phases be running to perform its task*—For example, since our powder blender required liquid ingredient flow in order to work, our *Add Cocoa* phase needed the *Add Sweetener* and *Add Water* phases to be running.

Figure 5.7 shows an expansion of the *Add Ingredients* operation from Figure 5.6. Each phase performs a unique and independent function.

Figure 5.7 Add Ingredients Operation

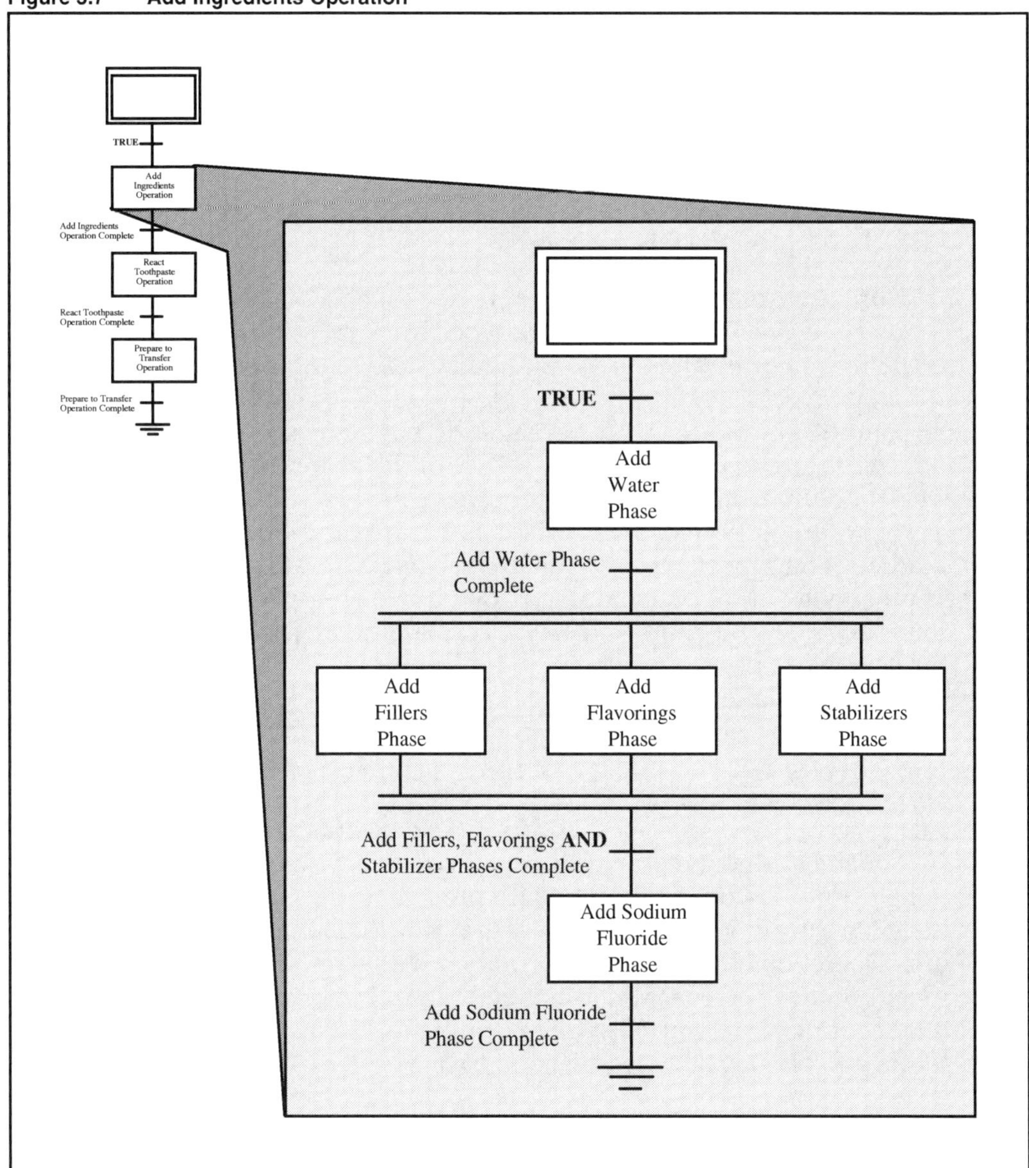

Phases and phase logic need to mimic the way your operators manually run the process (i.e., no computer assistance). For example, if someone has to stop a running procedure, what must be done to stop it efficiently and effectively?

By now, you're asking, "What's with this toothpaste example? Where's an ice cream example?" The answer is very simple. At Ben & Jerry's, all of the mixes are made starting at the operation level. That is, we only needed a single operation, consisting of several phases, to make a batch of mix—no procedures, no unit procedures. OpenBatch offers the option of starting a recipe at the procedure, unit procedure, or operation level, so we started all of our recipes at the operation level. (Technically, this violates S88, which states that all recipes must start at the procedure level. However, Sequencia saw a practical need for such a feature.) Since we wanted to provide an example that showed how to use all four levels of the procedural control model, we dreamed up this toothpaste batching procedure. But because we know you're interested, Figure 5.8 shows the basic idea behind making chocolate mix for ice cream.

Recipe Collapsibility

The procedural control model is designed to be "collapsible." For example, perhaps you have a recipe that requires a procedure and several unit procedures. However, your unit procedures are straightforward or simple enough so you have no need for operations, so you jump right to phases. This is allowed by S88, although some S88-aware software packages do not allow this explicit collapsing. With some packages, you cannot "skip" levels of the procedural control model. If you have a procedure and phases, you must have unit procedures and operations. Getting around this is simple. If you wish to use unit procedures and phases but not operations, you simply place all the phases in an operation and then "wrap" that operation with a unit procedure. We could have wrapped our single recipe operation into a unit procedure and wrapped that into a procedure, but we chose not to. Instead, we chose to use OpenBatch's feature for starting a recipe at the operation level.

Converting Site Recipes into Master Recipes

Creating a procedure in a master recipe from a procedure in a site recipe may be difficult. Remember that the process model (Process; Process Stage; Process Operation; Process Action) deals only with process functionality and not equipment. On the other hand, the procedural control model (Procedure; Unit Procedure; Operation; Phase) is tightly coupled with equipment, namely, units. A process stage in the process model is not tied to a unit, but a unit procedure in the procedural control model must be run to completion on a single unit. This means that a single process stage may correspond to one or more unit procedures.

Likewise, process operations in the process model are not limited to a unit, but an operation in the procedural control model must run to completion on a single unit. Therefore, a single process operation might require more than one operation to perform the processing desired. Things get even trickier at the lowest level of both models. One process action from the process model may require several phases from the procedural control model in order to carry out the processing needed. However, several process actions may be accomplished by a single phase.

Figure 5.8 A Typical Operation for Making Chocolate Mix

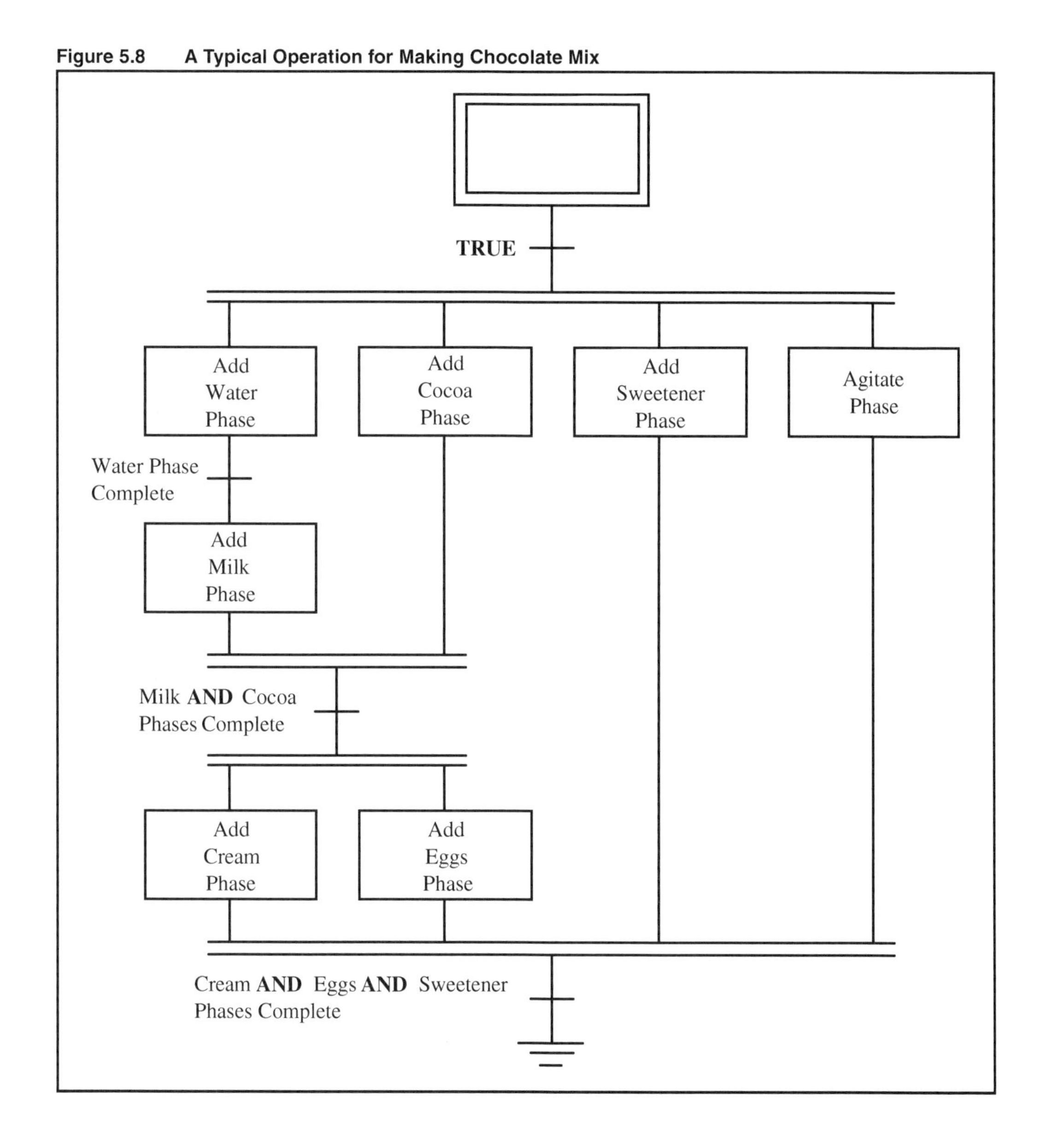

Linking the Physical, Procedural Control, and Process Models

There is a link between the procedural control model and the process model. However, the link also involves the physical model. It's simple and goes like this: "To accomplish process functionality, you need equipment and a procedure to control that equipment." In other words, the process that is required to make a batch of product is defined with the process model. To carry out that batch process, you need equipment, as defined by the physical model, and a procedure to control that equipment, as defined by the procedural control model. Figure 5.9 shows the relationship between the three models.

Figure 5.9 Procedural Control/Equipment Relationships Necessary to Achieve Process Functionality

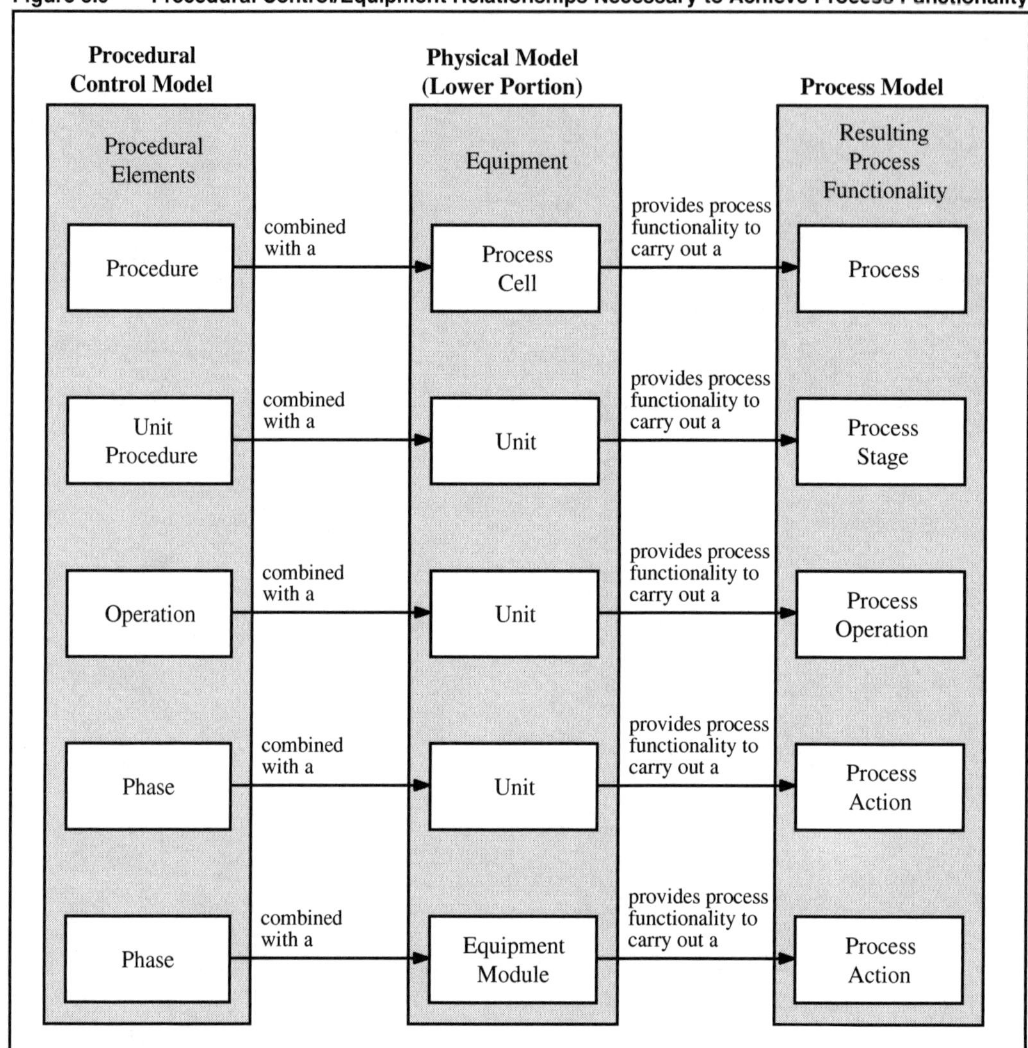

Notice how units can run unit procedures, operations, and phases to accomplish process stages, process operations, or process actions. Also notice that a phase can run against a unit or an equipment module to accomplish a process action. All of the phases for making ice cream mix, like *Add Water* or *Agitate*, are executed by a unit (one of the batch tanks), not by equipment modules. This is because the version of OpenBatch we used only allowed units to execute phases, not equipment modules. However, the software you are using or plan to use (including newer versions of OpenBatch) may allow equipment modules to execute phases. This can be a very powerful feature. Use it if you have the opportunity.

Now let's learn about the other parts of a recipe.

6

Recipes, Part 2: All the Other Stuff

It's a recipe that tells us how to use equipment to combine ingredients (raw materials) and make a product. We learned in the last chapter that S88 defines four different types of recipes and five categories of information contained in a recipe. One category is recipe procedures. This chapter deals with the other four categories: the header, equipment requirements, formula, and other information.

Information in a Recipe

The information contained in a recipe varies for each type of recipe. Figure 6.1 is a reminder for you of the four recipe types discussed in Chapter 5.

Figure 6.1 S88 Recipe Types

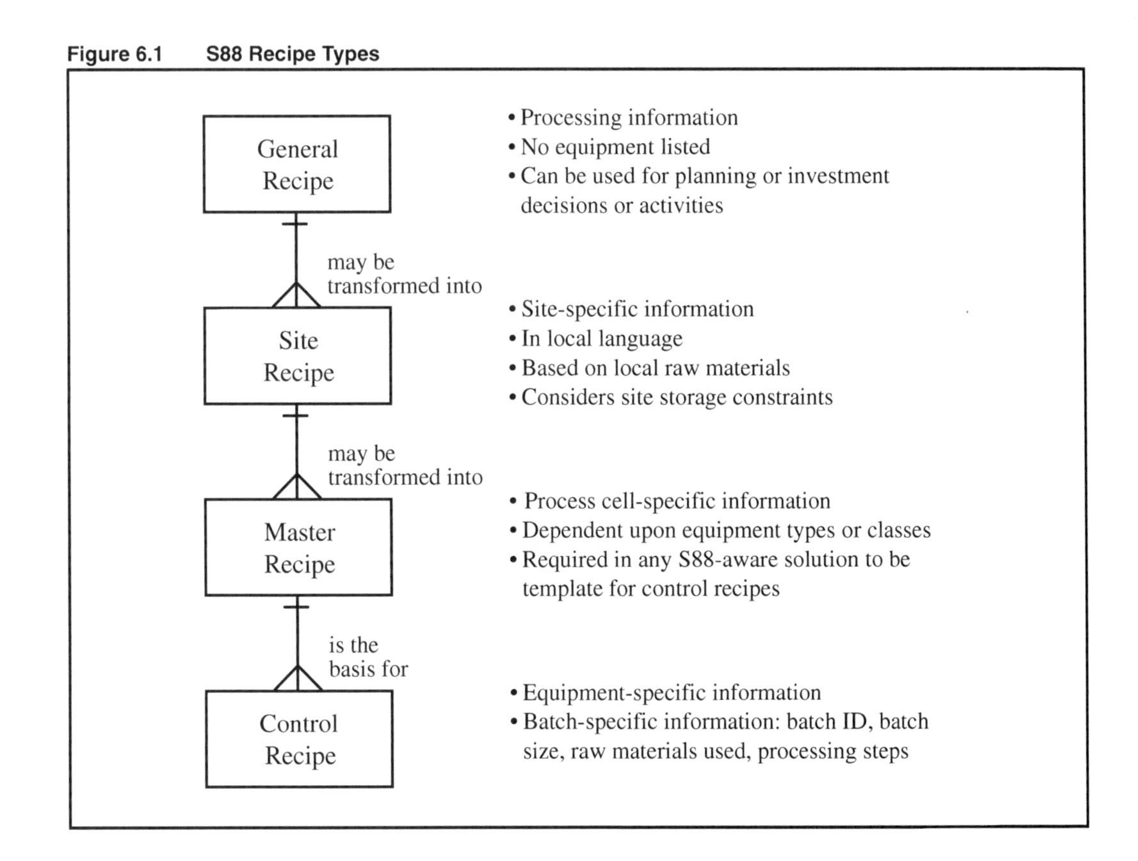

A recipe at any level (general, site, master, or control) contains five categories of information:

- Header
- Equipment Requirements
- Procedure
- Formula
- Other information

Category 1: The Header

The header includes both administrative information and a process summary. The administrative information may include recipe and product identification, the recipe version, the recipe originator (or author), the issue date, a revision history, approvals, and status (active, superseded, retired, etc.). The process summary briefly describes the manufacturing process.

Note that in all five categories, the information may change depending on the recipe level. For example, a site recipe header may also include the general recipe name and version from which it was derived.

Category 2: The Equipment Requirements

The equipment category provides information about the specific equipment needed to make a batch or a specific part of the batch. Since general and site recipes don't usually call out specific equipment, the requirements in these recipes are typically described in general terms. As we mentioned when discussing general recipes in Chapter 5, examples of these terms include materials of construction (e.g., a vessel of stainless steel versus one of galvanized steel) or processing characteristics (e.g., maximum mixing shear or pressure range).

At the master recipe level, the equipment category uses the guidance of the general or site recipes to call out classes of equipment, such as a pressure-controlled vessel or high-shear mixer. The control recipe equipment category identifies specific equipment, such as Mix Tank 1.

Category 3: The Procedure

As you learned in Chapter 5, a recipe procedure defines the strategy for carrying out a process. That is, the procedure explains how ingredients (raw materials) are to be combined, reacted, or otherwise processed to make a batch. The procedure describes a structured sequence of processing activities that are necessary to make a product.

There are two different models used to define recipe procedures, depending on the recipe level: the process model and the procedural control model. You were introduced to both of them in Chapter 5. Since the general and site recipes focus on processing activities rather than on specific equipment, the procedures for the top two recipe levels are based on the process model. But a transformation occurs between the site and master recipe levels. Master and control recipes deal with equipment classes or specific pieces of equipment. That's why these recipes are based on the procedural control model. Figure 6.2 shows both models.

Note that for general and site recipes the top level is *Recipe Procedure*—not *Recipe Process*. This is how the S88 committee defined the model. Here's our best guess as to why: even though the general and site recipes are based on the process model, the defined sequence of execution is still called a recipe *procedure*.

Category 4: The Formula

The formula describes recipe process inputs, process parameters, and process outputs. This is shown conceptually in Figure 6.3.

Figure 6.2 Process and Procedural Control Models for Recipes

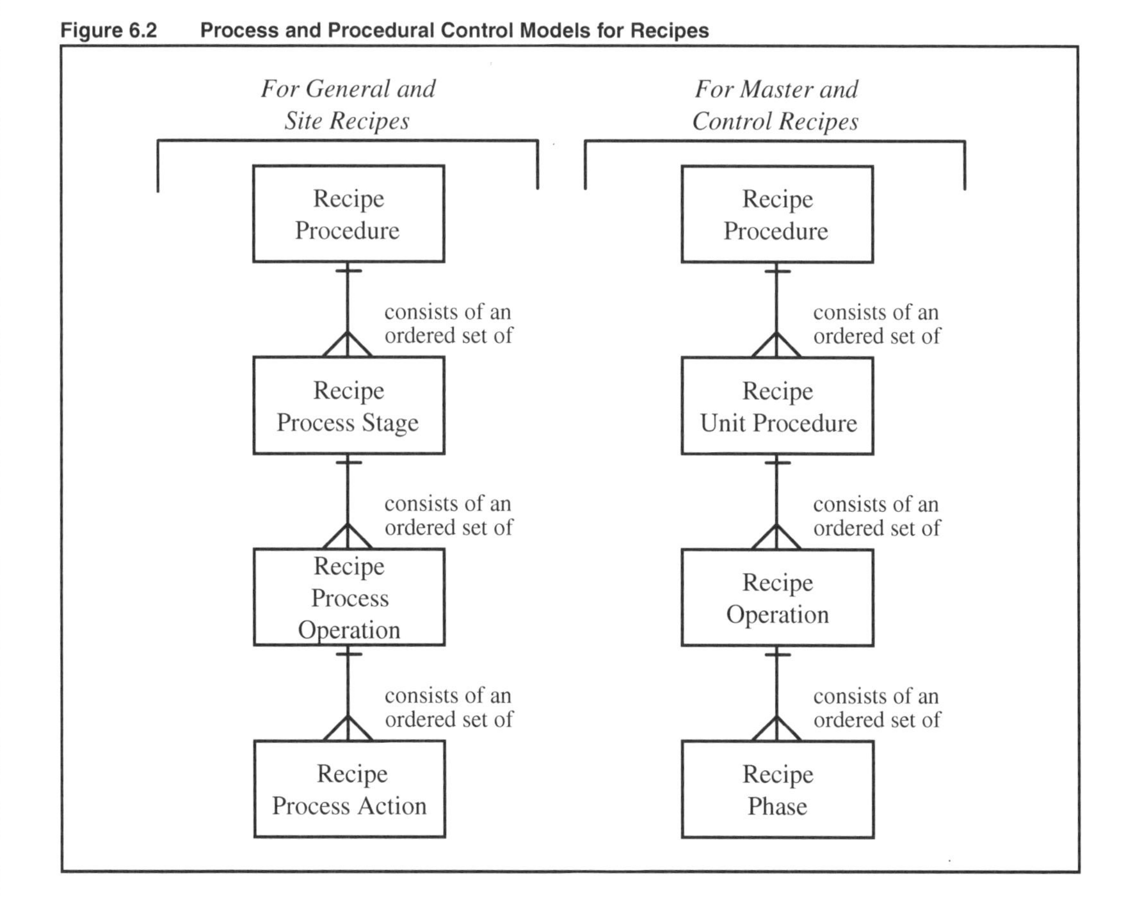

Figure 6.3 Recipe Formula Information

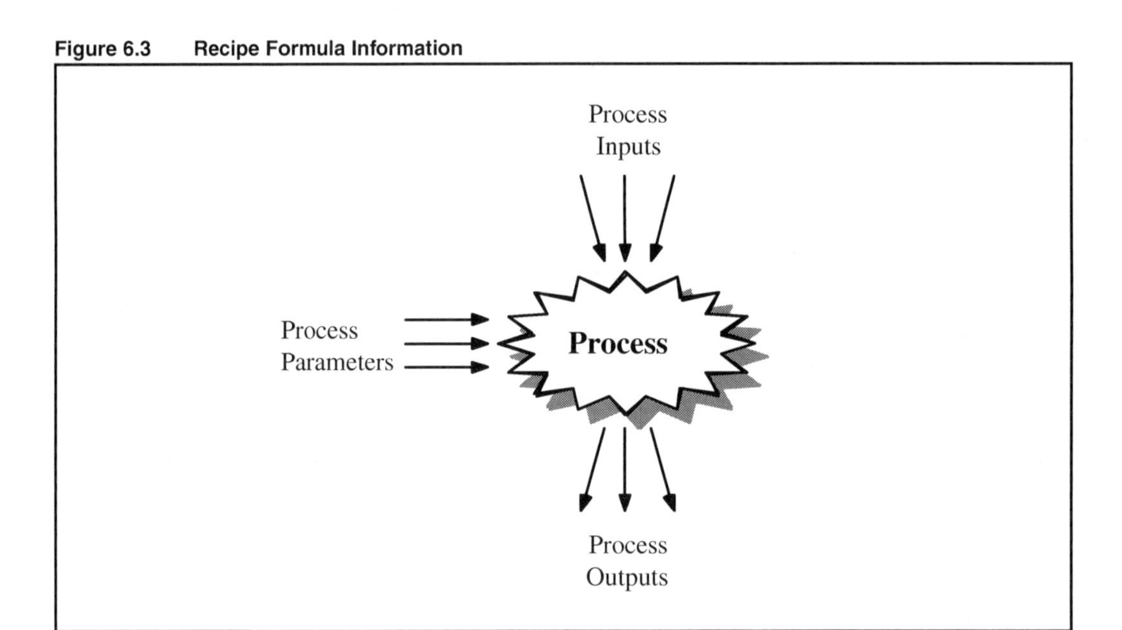

Process inputs include the raw materials and quantities required to make a batch of product. Not all raw materials may be included in the final product; it depends on your process. For example, a catalyst that is later recovered or stripped can be a raw material. Process inputs can also include energy (electricity, steam, etc.) or other resources, such as human labor. Note that all quantities in formulas may be absolute, or they may be formed by calculations based on the expected batch size.

Process parameters provide information such as reaction temperature, ingredient transfer rate, mixing speed, or holding time. Process parameters can be measured values, such as ingredient quantities, or attributes, such as "Powder Incorporation ON/OFF." Parameters can be used as set points in PID loops, as comparison values, or in conditional logic. In our recipes at Ben & Jerry's, process parameters included raw ingredient sources, specific ingredient flow paths, and agitator speed.

A *process output* identifies the material (product) and its quantity that result from executing a single batch. The process output called out in each of our recipes is a specific mix type and an expected quantity. Process outputs might also include by-products or recovered materials, such as catalysts. This recipe category may include yield information or data concerning the environmental impact of the recipe product or by-products.

Formulas can be used to distinguish between different products or different product "grades" that are defined by the same procedure. For example, in recipes that include dairy ingredients, each type of product can use one recipe (and procedure), regardless of the varying potency values of raw dairy. To create batches that meet recipe specifications use recipe formulas to combine different amounts of dairy ingredients and water.

There are many ways to implement formulas in a batch control system. OpenBatch has downloadable elements called *parameters*, but according to S88 definitions not all of these parameters are truly process parameters. Some can be classified as process inputs, such as ingredient types and quantities. OpenBatch allows for essentially any number of downloadable parameters. However, each parameter is a tag (see Chapter 4 for a definition of "tag"). And remember, a tag equates to money.

At Ben & Jerry's, we chose to use OpenBatch parameters for items that would not necessarily change from batch to batch, such as agitator speeds. We used our human-machine interface (HMI) to gather potency information about our raw tanks, accept batch size input from a mix maker, calculate recipe ingredient quantities, and download that information to a PLC.

Organize the formula values in a recipe so they are best suited to the way you run your batching systems. You may choose to have a master formula list for the entire recipe procedure, or you may wish to have separate lists for each recipe procedural element. Either way is fine. Do what's best for you.

Category 5: Other Information

The final information category in a recipe is generalized as "other information." Often this means comments about the recipe. These can include recipe-dependent safety comments, but comments are not intended to be a substitute for the information in a material safety data sheet (MSDS). You should not include comments about ingredients (raw materials) in a recipe. Ingredient comments (like odor and color) should be kept in an ingredient specification. The comments should focus on the recipe itself.

Special data collection or reporting requirements can also be listed in "Other Information." These comments should call out particular data collection or reporting activities not normally performed for a typical recipe. Recipe-dependent compliance comments (such as extra activities surrounding packaging or labeling information) might be listed in this section as well.

How recipe information appears is unimportant. Some companies prefer paper copies with a specific format. Others wish to keep everything electronic and transfer data using a secure, controlled procedure. S88 only suggests the five sections discussed in this chapter; it really doesn't care how they look.

Well, that's recipes. Whew! Now we need to discuss how recipes are linked to equipment control. That's one of the subjects of Chapter 7.

7

Linking Recipes to Equipment

So far, we've discussed equipment and recipes using the physical model, the procedural control model, and the process model. Chapter 4 was all about equipment, and Chapter 5 showed how recipes defined the batching process using procedures. But somehow recipe procedures must be linked to equipment control. An operator can make an entire batch by controlling equipment one function at a time, but he or she assumes responsibility for all kinds of details, such as coordinating equipment control and allocating equipment. In other words, to make a batch without requiring incredible intervention on the part of the operator a batch management system must be used. That's what this chapter is all about. Don't feel overwhelmed when you first read this material. Some of this stuff is fairly intense. Give it a little time to sink in.

Types of Control

Before we can talk about linking recipes to equipment control, we really should discuss the different types of equipment control common to batch manufacturing regardless of the level of automation (if any) used. In the world of S88, there are three types of control: basic control, procedural control, and coordination control.

S88.01 defines *basic control* as "the control dedicated to establishing and maintaining a specific state of equipment and process." The standard includes such items as the following in the realm of basic control:

- *Regulatory control*—This is not the control exercised by OSHA, the EPA, the FDA, and other joyous, fun-loving quasi-legislative agencies. It's a system of control that attempts to maintain one or more process variables at or near a desired value. Perhaps you wish to regulate the flow rate of a raw material during batching, or maybe the pH of your clean-in-place (CIP) detergent must be within a specified range.
- *Interlocking*—Use this form of basic control to prevent a function from starting or to stop or hold a function. For example, many companies use flow panels to make connections for routing liquid flow. Proximity switches on these panels can be linked to basic control logic, thereby preventing functions from starting unless a "prox" switch shows a swing in place. Process variables, such as temperature, flow rate, or pH, can also serve as conditional interlocks.

- *Monitoring*—Basic control can monitor a process and trigger an alarm if a variable falls out of a specified range.
- *Exception handling*—Taking alarming one step further, basic control can enable an exception-handling function. For example, if a flowmeter indicates a no-flow condition, this basic control can request that a batch be placed in *Hold*.
- *Repetitive discrete or sequential control*—This is similar to regulatory control but is oriented toward controlling a given state (such as *Open* or *Closed*) rather than a given value. A basic stop valve controlling flow is one example, but discrete (or sequential) control can use more than two states. A three- or four-way valve that dispenses ingredients is also an example of discrete control.

Basic control may be activated, deactivated, or modified by operator commands or by procedural or coordination control. All levels of the physical model relevant to S88 can execute basic control: process cells, units, equipment modules, and control modules. Here are some other examples of basic control:

- *Stand-alone control functions that run equipment but don't interact directly with other control functions*—However, the controlled equipment must operate in order for some other function to perform correctly. For example, boiler control can be basic control. At Ben & Jerry's, our pasteurizer and CIP systems required steam, but our PLCs did not explicitly tell the boiler control system to produce more steam. Instead, the boiler control system monitored the steam load and adjusted settings accordingly.
- *Manual control of equipment*—Maybe you call this "operator overrides." We used the terms *force on* or *force off*, implying that regardless of what the controller is commanding a device to do the operator is forcing it into a particular state. (Note that we're not talking about a PLC force but an operator-initiated force from an HMI.) We'll discuss this further in the section on modes in Chapter 8.

The second type of control, *procedural control*, directs equipment-oriented actions to take place in a given sequence in order to carry out a process-oriented task. Déjà vu, you say? Absolutely. Chapter 5 was all about recipe procedures. To save trees, we're not going to repeat all of that information here. But keep in mind one important note: procedural control is a key characteristic of batch processes. It's the type of control that enables equipment to make a batch.

Figure 7.1 shows a matrix of the elements of procedural control that can be executed by the different levels of equipment. Since S88 was designed for implementing any level of automation, we're not making any assumptions that the procedural control execution is being handled electronically. More than likely, you'll use S88 software to run your process, but an operator can also execute a manual procedure.

Since a procedure spans units, only a process cell can execute it. A unit has the most opportunity to execute procedural control. An equipment module is confined to only executing a phase, and a control module is not allowed to execute any procedural control.

Figure 7.1 Procedural Control Executed by Equipment

	Procedure	Unit Procedure	Operation	Phase
Process Cell	◆			
Unit		◆	◆	◆
Equipment Module				◆
Control Module				

What differentiates an equipment module from a control module? According to some members of the standards committee, it's simply that an equipment module can execute a phase and a control module can't. In other words, when you're considering whether to place a procedure function in an equipment module or control module, your choice depends on whether the module needs to execute a phase.

According to S88, *coordination control* "directs, initiates and/or modifies the execution of procedural control and the utilization of equipment entities." Like basic control, coordination control handles the fundamental functions necessary to carry out a process—in our case, making a batch. But coordination control takes care of the big-ticket items, such as the following:

- *Managing equipment resources*—Coordination control supervises the availability and capacity of equipment, coordinates common resources, allocates equipment to batches, and arbitrates requests for equipment allocation in case more than a single procedure needs the equipment simultaneously.
- *Selecting procedural elements to be executed*—Coordination control helps determine which procedural elements to run. For example, S88 "law" says a unit can run only one operation at a time. Perhaps within a unit procedure, two operations exist in parallel, but a status flag determines which operation can run. Coordination control will look at that status flag and start the appropriate operation.

- *Propagating modes*—If a procedural element experiences some type of failure that should place a batch in hold, it's coordination control that propagates the *Hold* state up and down the procedural chain. For example, if a phase detects a flow failure, it may place itself in *Hold*. Then coordination control propagates that *Hold* state to the phase's corresponding operation, unit procedure, and procedure. (Depending on your batch management software, you may have different options for controlling this propagation.) Another example of propagating modes would be that of an operator who executes a command to put a batch in *Hold*, where coordination control propagates that command from the procedure to all unit procedures, operations, and phases.

The further "down" the physical model you go, the less coordination control is typically needed. Major pieces of processing equipment, such as units, are usually shared. However, more specific pieces of equipment, like control modules, are generally dedicated to a specific function (but they don't have to be). Figure 7.2 describes this phenomenon.

Figure 7.2 Coordination Control with the Physical Model

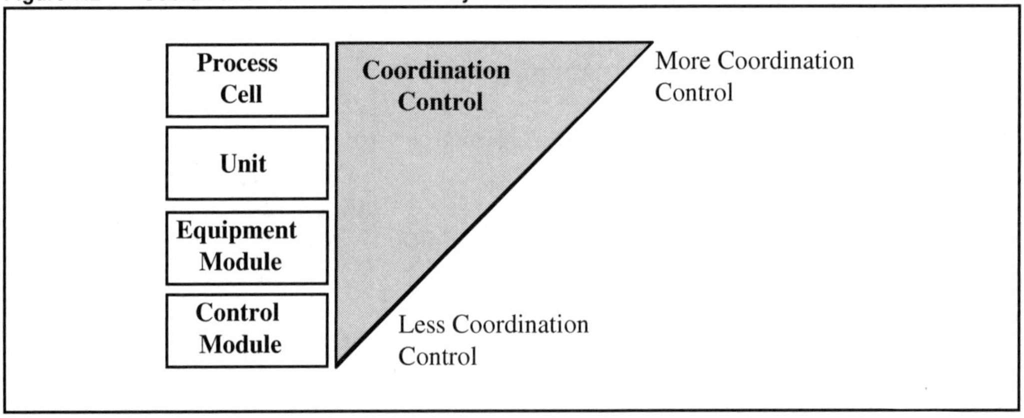

Chapter 9 discusses batch management and information functions, many of which fall within the realm of coordination control. So, we'll return to coordination control then. Now, enough of this background stuff . . .

Linking Recipes and Equipment Control

In order to make a batch, the instructions in a control recipe procedure must somehow be conveyed to the control system running the equipment. (Remember that a control recipe is the element of the recipe model that is specific to a particular batch.) Control recipes and equipment control can both have procedures, unit procedures, operations, and phases. Figure 7.3 shows both hierarchies.

Figure 7.3 Control Recipe and Equipment Control Procedures

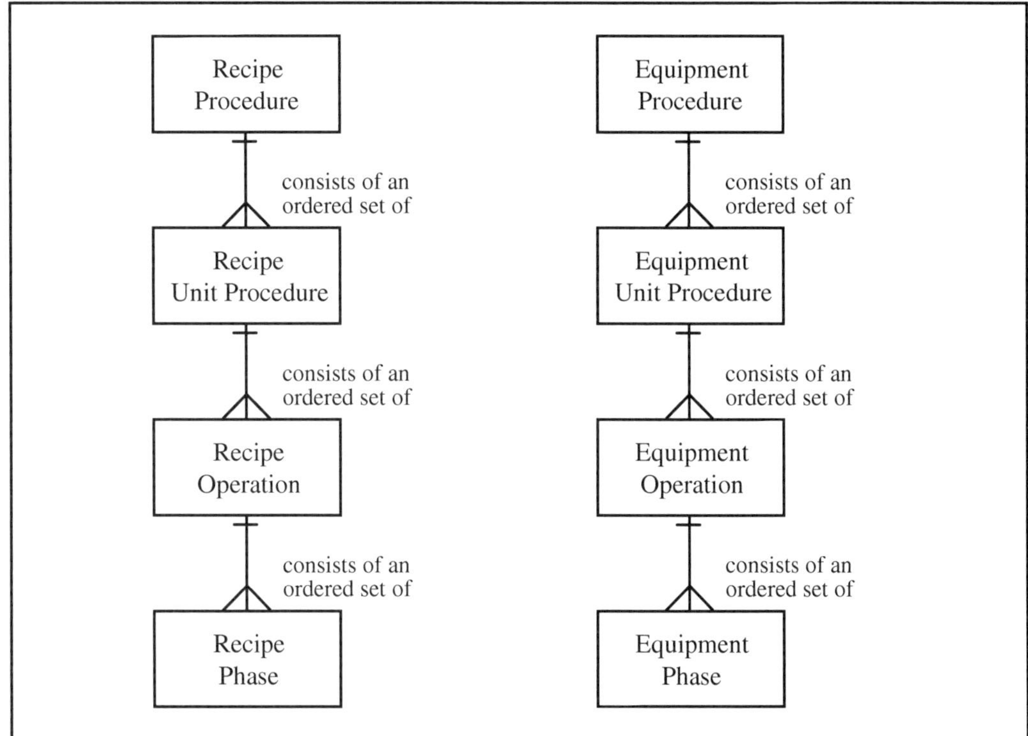

S88 suggests that a batch management system and an equipment control system should exist separately. To us, this means that the batch management system will handle elements of the control recipe procedure, and the equipment control system will handle elements of the equipment control procedure. While S88 does not exactly dictate where the link between the control recipe procedure and equipment control occurs, it does require that a control recipe procedure and an equipment phase always exist. Figure 7.4 shows this. Unit procedures and operations can either fall under the control recipe domain or the equipment control domain.

Figure 7.5 shows what has emerged as the most common link, the phase level link. At the phase level in the control recipe, a reference is made to an equipment phase that is executed in a control system so as to perform a particular batching function.

Linking at the phase level really means that the batch management system contains procedures, unit procedures, operations, and phases within the control recipe and that the equipment control system contains only phase logic. In this arrangement, an operator can only run individual phases using the control system. As we mentioned at the beginning of this chapter, the operator can make an entire batch in this fashion but assumes responsibility for details, such as coordinating phases and allocating equipment. Once more, to make a batch

Figure 7.4 The Procedural Control Model Links a Control Recipe Procedure with Equipment Control

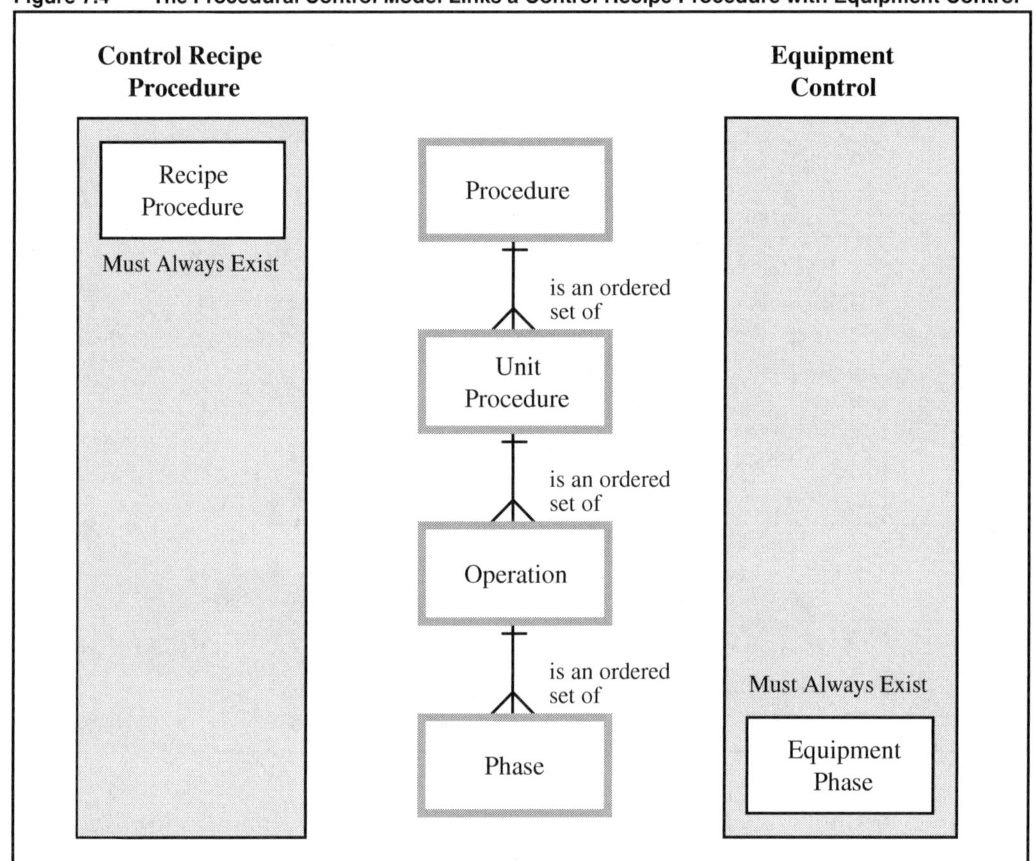

without requiring incredible intervention on the part of the operator, a batch management system must be used.

Of course, the linking does not have to be done at the phase level. Figures 7.6, 7.7, and 7.8 show linkages at the other levels.

OpenBatch and its derivatives (such as RSBatch, Total Plant Batch, and VisualBatch) use the phase level (Figure 7.5) to link control recipes to equipment control. Many pre-S88 batch control systems, especially those using PLCs, link the control recipe with equipment control at the procedure level, shown in Figure 7.8. This makes sense. Before the advent of PC or even minicomputer batch management systems, most batch management functions—including recipes—were handled on paper, and all the control was placed in a PLC.

Figure 7.5 Control Recipe and Equipment Control Linking at Phase Level

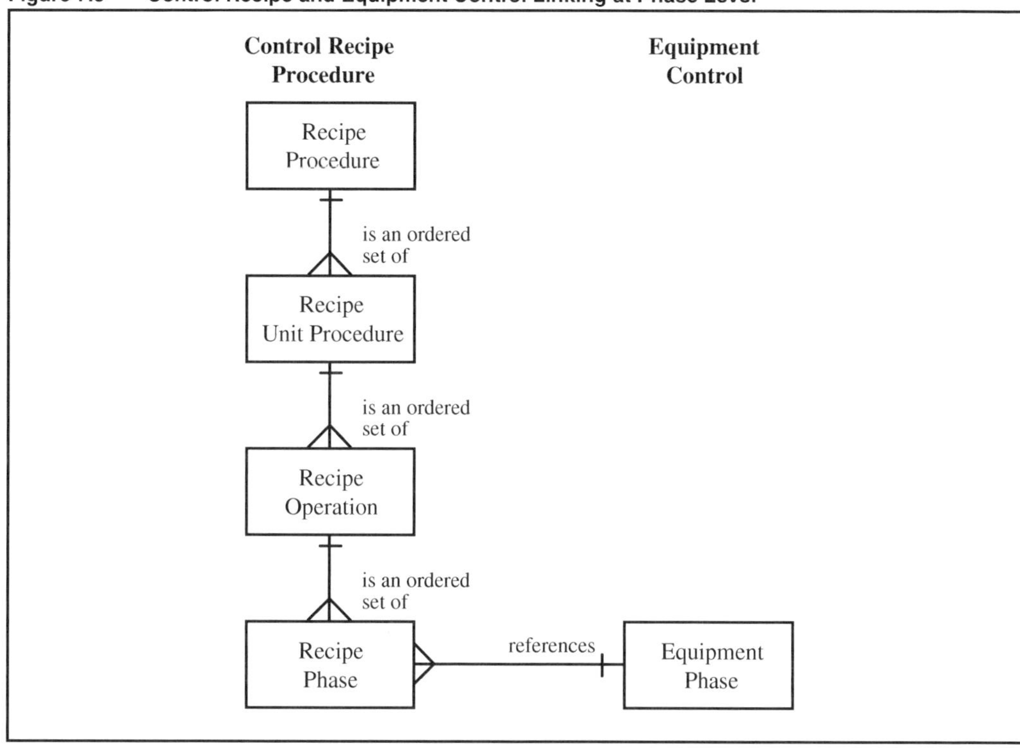

Figure 7.6 Control Recipe and Equipment Control Linking at Operation Level

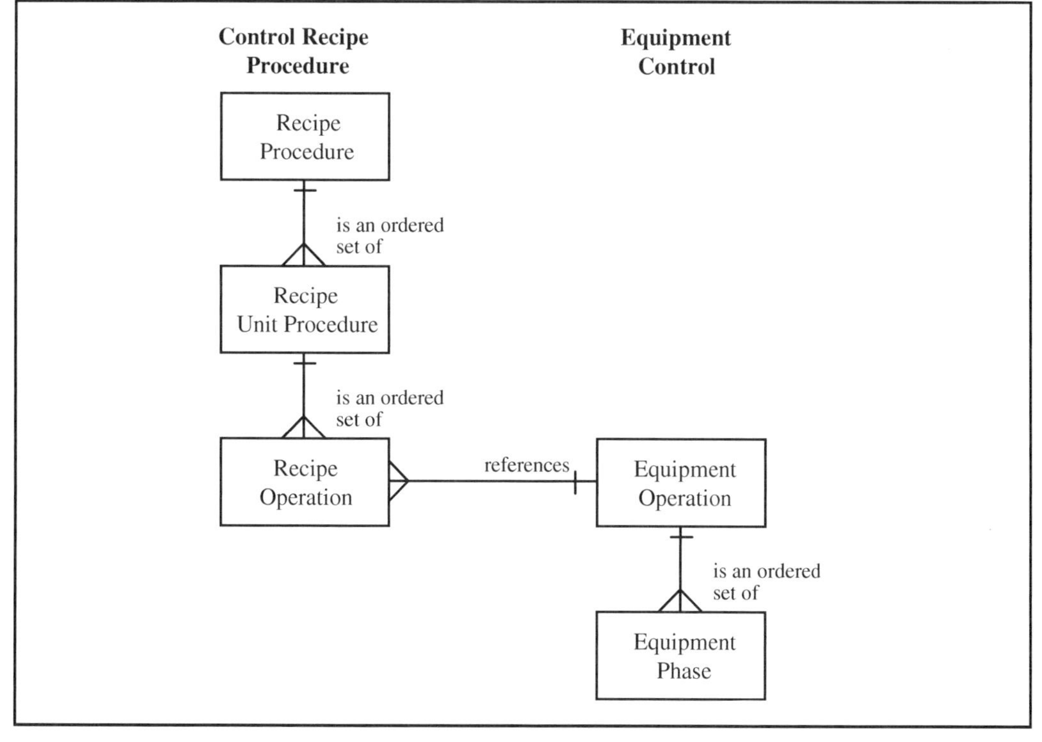

Figure 7.7 Control Recipe and Equipment Control Linking at Unit Procedure Level

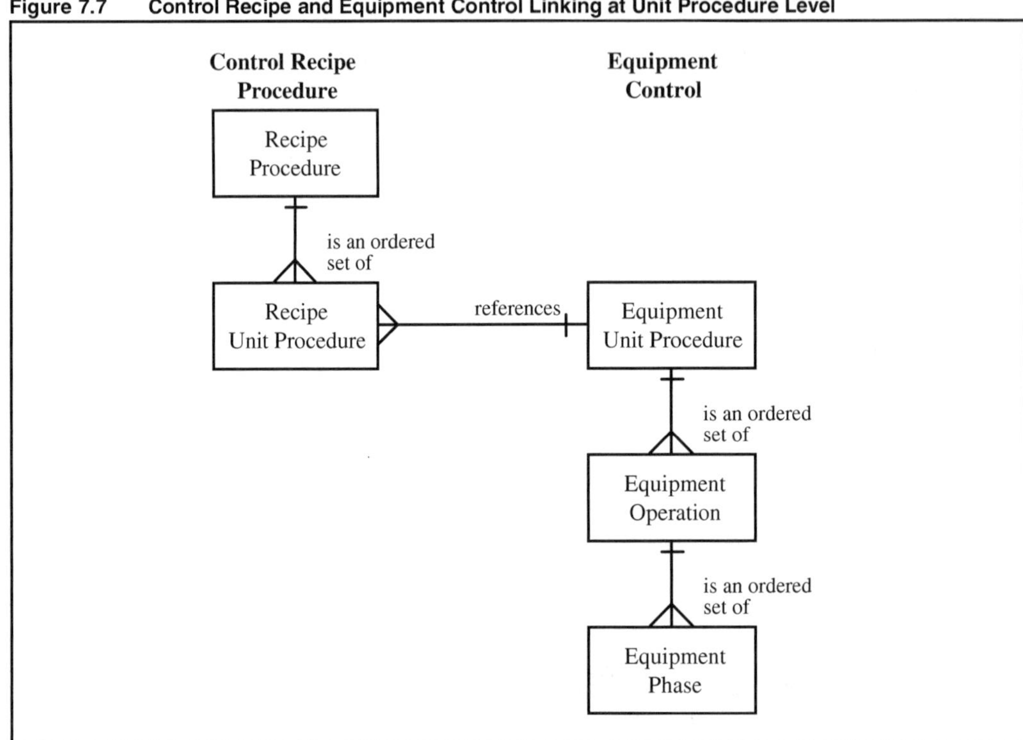

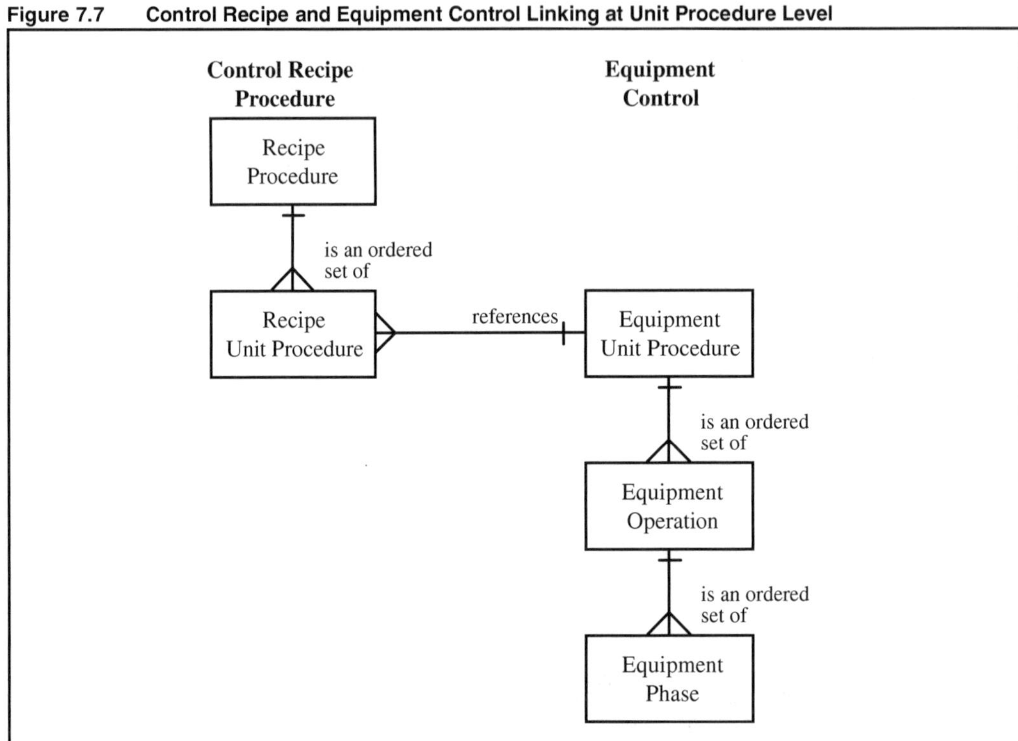

Figure 7.8 Control Recipe and Equipment Control Linking at Procedure Level

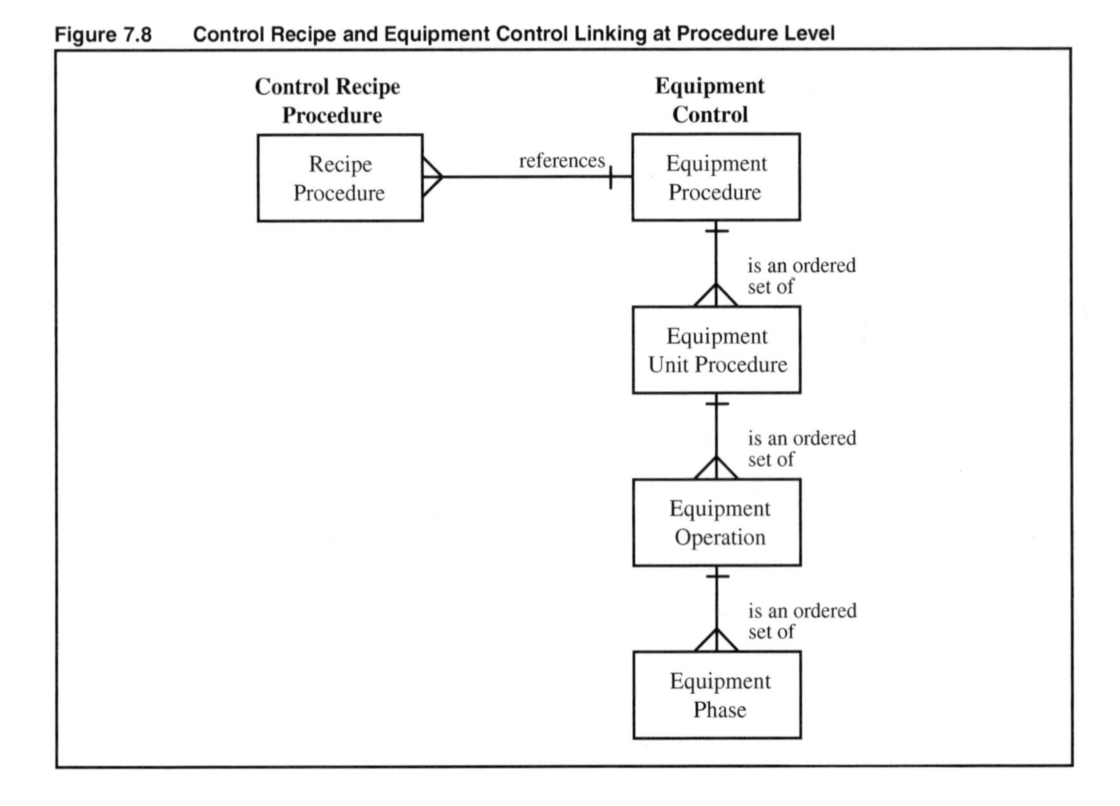

Back in Chapter 1, we explained that one of S88's great strengths was that it separated recipes from equipment control. Considering that traditional batch systems placed all of the procedural elements within equipment control, you can now see how difficult it was to develop, document, and maintain that code. Imagine placing the equivalent of several procedures and unit procedures, and potentially dozens of operations and phases, in a PLC. And now imagine writing the ladder logic required to not only perform all of those functions but also to coordinate which should occur when!

The point-and-click approach to desktop computing inherently makes maintaining procedures as part of the control recipe much easier. Isolating phases in a process-connected device, such as a PLC or DCS, provides modularity that is easy to develop, maintain, and replicate. To end users or vendors, it's easier to install and validate a batch control system and to duplicate one process cell or site to another. All this means time and money are saved developing a batch control system.

In Chapters 11 and 12, we'll talk more about equipment phases and writing phase logic. But first, we need to address some more important aspects of batch control.

8

Other Important Batch Control Items

To complete the puzzle on making batches using S88, we need to tie together what we've learned so far with some new stuff, to be introduced in this chapter. This new stuff includes—are you writing this down?—using *modes of operation*, understanding *states and commands associated with batch control*, dealing with control *exception handling* (abnormal conditions), and managing *allocating and arbitrating equipment use.*

Modes of Operation

A mode determines how *procedural elements* (such as procedures and unit procedures) and *equipment entities* (such as units and equipment modules) respond to commands and how they will operate. We all know about modes: automatic, manual, and so on. S88 formally defines a mode as "The manner in which the transitions of sequential functions are carried out within a procedural element or the accessibility for manipulating the states of equipment manually or by other types of control."

Modes affect procedural control and basic control. They don't really apply to coordination control. The S88 standard suggests three modes for procedural elements (*automatic, semiautomatic*, and *manual*) and two modes for equipment entities (*automatic* and *manual*).

For procedural control, the mode determines how the procedure progresses and who can affect that progression. In other words, the mode determines the way transitions behave. In automatic mode, transitions occur without interruption after the transition meets its conditions to fire. In semiautomatic mode, the procedure requires manual (operator) approval to proceed after the transition meets its conditions to fire. Transitions cannot be forced, but the standard allows procedural elements to be skipped or re-executed. In manual mode, the procedural elements and their order of execution are completely specified by the operator. Table 8.1 summarizes the modes for procedural elements.

Table 8-1. Possible Implementations of Suggested Modes for Procedural Elements

Mode	Behavior	Command
Automatic	Transitions within a procedure are carried out without interruption as necessary conditions are met.	Operators may pause the progression but may not force transitions.
Semiautomatic	Transitions within a procedure are carried out on manual commands as necessary conditions are met.	Operators may pause the progression and redirect the execution to an appropriate point, but operators may not force transitions.
Manual	Procedural elements within a procedure are executed in the order specified by an operator.	Operators may pause the progression and may force transitions.

For equipment entities, the mode determines what controls the function or who may manipulate a state. Let's use a control module as an example. A valve may have automatic and manual modes. In automatic mode, the valve could be under the control of a procedure or some control algorithm. In manual mode, the valve is under the control of an operator. Table 8.2 summarizes the modes for equipment entities.

Table 8-2. Possible Implementations of Suggested Modes for Equipment Entities

Mode	Behavior	Command
Automatic	Equipment entities are manipulated by a control algorithm.	Equipment entities cannot be controlled by an operator.
Manual	Equipment entities are not manipulated by a control algorithm.	Equipment entities can be controlled directly by an operator.

Procedural elements and equipment entities may change mode when a command is given by an operator or one is generated in another procedural element. A mode change can only occur when defined, required conditions for the change are met. Probably the most common way to change modes is with an operator command. Perhaps an operator wants to manually open a valve regardless of the state of the process. (Or Jim and Larry realize they opened the wrong valve and want to force it closed quickly . . .)

Consider what happens when a pasteurizer starts up—an example of a procedural element changing the mode. Normally, a basic form of control, such as a PID loop, controls the heating portion of a pasteurizer. Depending on how the loop is tuned, it may have great steady-state control of the process but slow response time on start-up. To quicken the start-up time, a phase controlling the pasteurizer may place the control module that is executing the PID loop in manual and force open a steam valve, ramping the pasteurizer heat faster. Once a desired temperature is reached, the phase returns the control module to automatic mode.

A mode change in one procedural element or equipment entity may force a change of mode in others. If an operator changes a unit procedure mode from automatic to semiautomatic, it may be wise for him or her to propagate all dependent operations and phases to semiautomatic mode. The S88 standard recognizes that propagation can move from a higher-level entity to a lower-level entity or vice versa, but it does not specify any propagation rules. Different batch management packages may have rules or may provide options for the end user.

States and Commands Associated with Batch Control

Procedural elements and equipment entities may have *states*. So can people: states of mind, a state of exhaustion, a state of undress. In regard to S88, the state is supposed to completely specify the current condition of a procedure element or equipment entity. *Commands* are one method for moving a procedural element or equipment entity from one state to another.

S88 suggests a common set of states and commands for procedural elements. Some committee members might argue that saying the standard *suggests* states and commands is too strong a statement; they'd prefer to say S88 "uses example states and commands." Regardless, the standard does not require any particular set of states and commands. However, keep in mind that the S88 committee members are pretty smart cookies. The example procedural states and commands they chose probably will work in more than 95 percent of all the batching systems used today. OpenBatch, VisualBatch, RSBatch, and Total Plant Batch all use the states and commands included in the standard. Table 8.3 is the state transition matrix for the procedural element states and commands suggested by the standard.

Table 8-3. State Transition Matrix for States and Commands Suggested by S88

		Command							
Initial State	**Next State if No Command Given**	**Start**	**Hold**	**Restart**	**Pause**	**Resume**	**Stop**	**Abort**	**Reset**
Idle		Running							
Running	Complete		Holding		Pausing		Stopping	Aborting	
Complete									Idle
Holding	Held						Stopping	Aborting	
Held				Restarting			Stopping	Aborting	
Restarting	Running		Holding				Stopping	Aborting	
Pausing	Paused		Holding				Stopping	Aborting	
Paused			Holding			Running	Stopping	Aborting	
Stopping	Stopped							Aborting	
Stopped								Aborting	Idle
Aborting	Aborted								
Aborted									Idle

Figure 8.1 shows a simplified state transition diagram for the states and commands as suggested by S88. What it doesn't show are all the commands from all the states. The diagram really only focuses on the *Idle*, *Running*, and *Complete* states. For example, if an operator issues a **HOLD** command, the batch will stop executing *Running* logic, transition to the *Holding* state, and begin executing *Holding* logic. According to Table 8.3, once the batch is in the *Holding* state, an operator can still issue **STOP** or **ABORT** commands. To keep the diagram from getting too cluttered, those commands from the *Holding* state aren't shown in Figure 8.1. (The same commands would also need to be shown from the *Held* and *Restarting* states. In addition, other lines need to be shown from the *Pausing*, *Paused*, *Stopping*, and *Stopped* states.)

Figure 8.1 State Transition Diagram for States and Commands Suggested by S88

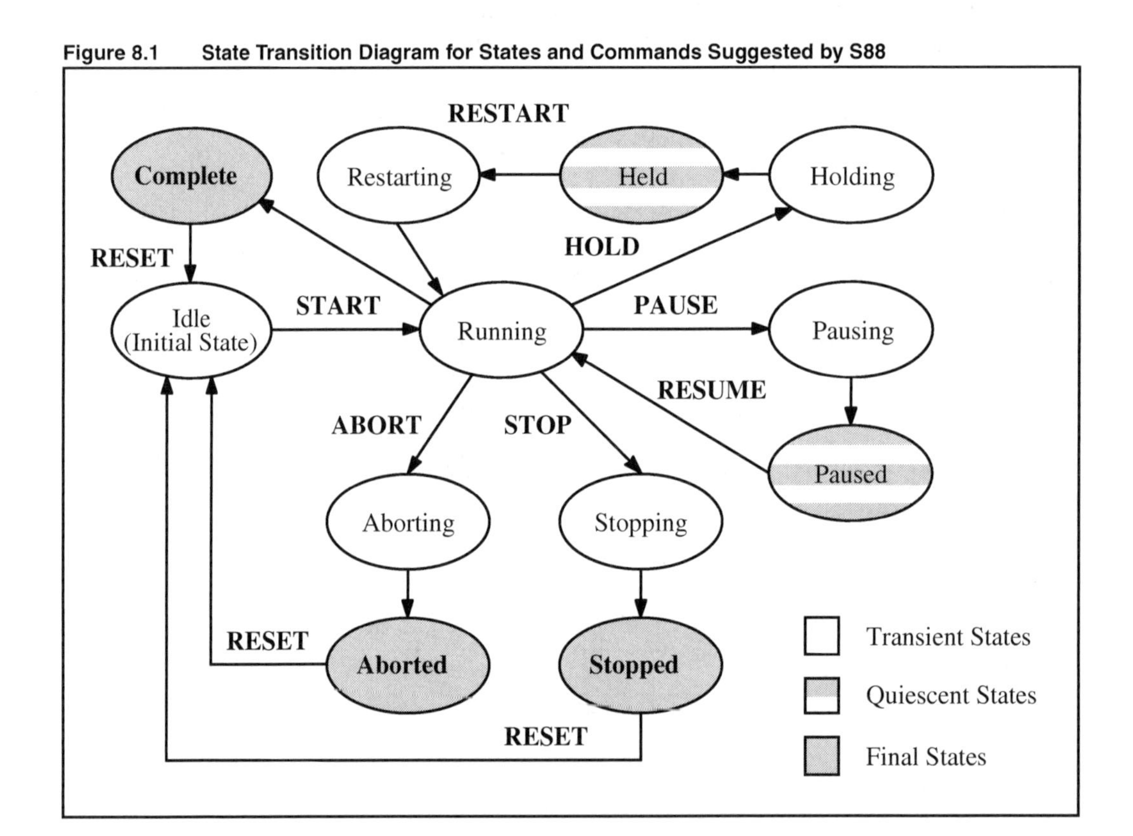

Table 8.4 describes each of the states. Table 8.5 describes each of the commands.

Table 8-4. Procedural States Suggested by S88

Idle	The procedural element is waiting for a **START** command that will cause a transition to the *Running* state.
Running	Normal operation.
Complete	Normal operation has run to a normal completion. The procedural element is now waiting for a **RESET** command that will cause a transition to the *Idle* state.
Holding	The procedural element has received a **HOLD** command and is executing its separate *Holding* logic to put the procedural element into a known condition. Once the *Holding* logic completes, the procedural element transitions automatically to the *Held* state. If no special sequencing is required to place the procedural element into a known condition, the procedural element transitions immediately to the *Held* state.
Held	The procedural element has completed its *Holding* logic and has been placed into a known or planned condition. The procedural element is now waiting for a command to proceed. This state is usually for longer-term batching interruptions.
Restarting	The procedural element has received a **RESTART** command while in the *Held* state and is executing its *Restarting* logic in order to return to the *Running* state. If no restarting sequencing is required, then the procedural element transitions immediately to the *Running* state.
Pausing	The procedural element has received a **PAUSE** command. This will cause the procedural element to stop at the next defined safe or stable location in its *Running* logic. Once the defined safe or stable location is reached, the state automatically transitions to *Paused.*
Paused	State reached once the procedural element has reached the next defined safe or stable location after a **PAUSE** command. A **RESUME** command causes a transition to the *Running* state, resuming normal operation immediately following the defined safe or stable location. This state is usually for shorter-term batching interruptions.
Stopping	The procedural element has received a **STOP** command and is executing its *Stopping* logic, which sequences a controlled, normal stop. If no stopping sequencing is required, then the procedural element transitions immediately to the *Stopped* state.
Stopped	The procedural element has completed its *Stopping* logic and is waiting for a **RESET** command to transition to the *Idle* state.
Aborting	The procedural element has received an **ABORT** command and is executing its *Aborting* logic, which sequences a quicker, but not necessarily controlled, abnormal stop. If no aborting sequencing is required, then the procedural element transitions immediately to the *Aborted* state.
Aborted	The procedural element has completed its *Aborting* logic and is waiting for a **RESET** command to transition to the *Idle* state.

Table 8-5. Procedural Commands Suggested by S88

Start	Orders the procedural element to execute its normal *Running* logic. Only valid while the procedural element is in the *Idle* state.
Hold	Orders the procedural element to execute its *Holding* logic. Only valid while the procedural element is in the *Running, Pausing, Paused,* or *Restarting* states.
Restart	Orders the procedural element to execute its *Restarting* logic to safely return to the *Running* state. Only valid while the procedural element is in the *Held* state.
Pause	Orders the procedural element to pause at the next programmed pause transition within its normal *Running* logic and await a **RESUME** command before proceeding. Only valid in the *Running* state.
Resume	Orders the procedural element to resume execution in its normal *Running* logic after it has paused at a programmed transition as a result of either a **PAUSE** command or a *Semiautomatic* mode. Only valid in the *Paused* state.
Stop	Orders the procedural element to execute its *Stopping* logic. Only valid while the procedural element is in the *Running, Pausing, Paused, Holding, Held,* or *Restarting* states.
Abort	Orders the procedural element to execute its *Aborting* logic. Valid while the procedural element is in every state *except Idle, Complete, Aborting,* and *Aborted.*
Reset	Orders the procedural element to transition to the *Idle* state. Only valid while the procedural element is in the *Complete, Stopped,* and *Aborted* states.

Remember that many batch control packages make the link between recipe procedures and equipment control at the phase level. The "-ing" states (*Running, Holding, Restarting, Stopping,* and *Aborting*) generally are controlled by separate blocks of code in a phase. For example, in a PLC each phase may be contained within a single file but have separate sections of ladder logic in that file for controlling running, holding, restarting, stopping, and aborting. We'll talk more about this in Chapters 10 and 11.

Notice that according to S88 there is a big difference between *Holding* and *Pausing*. When an operator (or procedural element) issues a **HOLD** command, a separate section of code begins executing (the *Holding* logic). When someone or something issues a **PAUSE** command, the *Running* logic simply pauses at the next appropriate location. When a **RESTART** command is issued after the batch enters the *Hold* state, still another section of code (*Restarting* logic) begins executing. When an operator issues a **RESUME** command after the batch enters the *Paused* state, the *Running* logic simply starts again at the location it paused at. You can do really neat (and really important) stuff with *Holding* and *Restarting* logic. We'll get into that in Chapters 11 and 12.

Like modes, procedural elements and equipment entities may change state when a command is given by an operator or one is generated in another procedural element. A state change can only occur when defined, required conditions for the change are met. Probably the most common way to change states is with an operator command.

Let's use a pumping control module as an example of an equipment entity changing the state. A couple of valves, a pump, and a flowmeter make up this control module. Let's say a phase issues a command to start charging an ingredient. A valve or two may open, the pump should start, and the flowmeter should soon start registering flow. If the flowmeter does not register flow within a given amount of time (maybe the pump motor tripped, a valve did not open, or a tank was empty), a failure flag will be set and a **HOLD** command automatically issued to the phase.

Also, like modes, a state change in one procedural element or equipment entity may force a change in state in others. If an operator issues a **PAUSE** command to a procedure, it may be wise for him or her to transition all dependent unit procedures, operations, and phases to the *Pausing* state. Likewise, in our flow failure example, a **HOLD** command sent to one phase might need to propagate up to the procedure level and then back down to all dependent procedural elements. Sometimes this is for the convenience of the operators so a problem can be fixed. Sometimes this is absolutely necessary for safety considerations: many processes require that ingredients charge simultaneously or at a proportional rate to each other to prevent a dangerous (or sometimes explosive) condition. Many of the batch management packages provide you with options regarding how state propagation should occur.

Looking back at Table 8.3, you can see the inherent priorities of the commands suggested by the S88 standard. The **ABORT** command has the highest priority because it can be issued from any active state (i.e., not *Idle* or *Complete*) except for the *Aborting* or *Aborted* states. Next highest in priority is **STOP**, then **HOLD**. Among the others, **RUN**, **RESTART**, **PAUSE,** and **RESUME** all have the same sort of priority because each can only be issued from one state.

States for equipment entities may be very different than states for procedural elements. For example, the states for a pump could include *on*, *off*, *percent on (or speed)*, *failed*, and *ramping*. The states for a valve could include *open*, *closed*, *percent open*, *failed*, and *traveling*. For some reason, the standard committee decided not to suggest a formal example set of equipment entity states. Maybe they were tired from a long debate on the procedural states.

Exception Handling

Wouldn't it be nice if all batches ran perfectly from start to finish? Well, we have some bad news to break to you: batches just don't behave as we would hope every time. An event that occurs outside the normal or desired behavior of making a batch is commonly called an *exception*. Handling these exceptions is an essential function of batch manufacturing and typically accounts for a large portion of the control definition. Don't be surprised if more than half of your design work and program code deals with exceptions.

Exceptions don't have to be related to failures like a tripped pump, stuck valve, or lack of steam. A tank running out of an ingredient in the middle of a batch could be considered an exception. Exception handling can occur in procedural, basic, or coordination control.

A response to an exception may cause a change in the mode or state of procedural elements or equipment entities. Using our flow failure example from the last section, a charge phase will detect an exception if the flowmeter doesn't register any flow. This causes a **HOLD** command to be issued to the phase, causing in turn a change in state from *Running* to *Holding* and eventually to *Held*.

Exception handling can be done in procedures or in equipment control. Procedure exception handling should be kept in the recipe because equipment control generally does not have any understanding of the procedures coordinating it. For example, if an operation has enough information to know that phase B and not phase A should start, handle that in the procedure, not in the phases.

You can also perform exception handling even before a recipe starts. Perhaps a recipe needs cream as an ingredient, but no cream tanks are available when the recipe starts. You can implement the exception handling in PLC or DCS phase logic. When the exception occurs, the equipment phase can issue a **HOLD** command to the batch. However, it may be better if you performed some quick checks via your HMI to see that cream is available before the batch starts.

Exception handling in equipment control is probably best handled directly in phase logic or in code that manipulates equipment or control modules. We've beat the flow failure example enough already, so consider the following example. If an ingredient can be stored in multiple tanks, you may wish to designate a primary and secondary ingredient tank. (Perhaps the operator can modify this selection.) When the phase that charges that ingredient starts, it pulls the ingredient from the primary tank. An exception occurs when the primary tank empties before the phase transfers the total amount needed. A control module monitoring the flowmeter will trigger an exception, causing the phase logic to switch to the secondary tank. This can all be done without operator intervention or holding or otherwise interrupting the batch. If the primary tank runs out and a secondary tank has not been selected, you may wish to prompt the operator for a secondary tank during the batch. Depending on your process or the volatility of your ingredients, you may not want to interrupt the batch.

Your batch management package may have recommendations regarding handling exceptions. Sequencia strongly recommends against manipulating equipment phase states directly. OpenBatch relies on an "executive" called a programmable logic interface (PLI) to monitor and maintain the states of all phases. Changing the states directly in equipment phases may confuse the poor PLI, which in turn may confuse the operators, which in turn may give you a headache.

To help manage state changes, OpenBatch allows for user-defined failures that can trigger different state changes. In the case of our flow failure example, the control module monitoring the flowmeter can set a failure flag associated with that phase. The phase logic can monitor that flag and react accordingly. At Ben & Jerry's, we created some phases that trigger a **HOLD** command if flow failures occur.

Allocating and Arbitrating Equipment Use

If batch processing equipment were really cheap, companies would purchase a whole lot of it, and no piece of equipment would be needed outside of any particular equipment train. But batch processing equipment is not cheap, so companies may have to share equipment for different batches. Sometimes equipment breaks down too, which makes it vital that you have substitute equipment on hand. Here's where *allocation* and *arbitration* come in.

As a particular batch or unit needs equipment and other resources to complete or continue processing, those resources must be assigned to it. Allocation is a form of coordination control that makes these assignments. When more than one batch or unit needs the same equipment or resource at the same time, arbitration determines who wins.

The S88 section on allocation says:

> The very nature of batch processing requires that many asynchronous activities take place in relative isolation from each other with periodic points of synchronization. Many factors, both expected and unexpected, can affect the time required by one or more of the asynchronous activities from one point of synchronization to the next. For those reasons, and because of the inherent variation in any manufacturing process, the exact equipment which will be available at the time it is needed is very difficult to predict over a significant period of time.

Wow, what a statement! A recipe does not have to be very complex to require different activities to take place separately (*asynchronously*). This is fine initially, but the resulting intermediate products must sometime later combine (*synchronously*) to produce another intermediate or a final product. When other recipes or unexpected maintenance make certain equipment unavailable, substitutes may be obtainable.

Your site may use sophisticated scheduling to try to optimize a recipe's processing sequence from the perspective of equipment usage. However, for those just-in-case situations you may wish to allow alternate equipment to be used if necessary. Recall from Chapter 4 that the path describes the usage and routing of the equipment that is necessary to make a batch. Allocation determines the path, whether fixed (the same equipment batch after batch) or dynamic (different equipment based on availability).

If more than one unit can request another resource, the resource is considered *common*. Common resources help reduce capital or operating costs while maximizing process flexibility. An expensive powder blender that serves two units is a great example of a common resource.

In S88 schemes, common resources are often implemented as equipment or control modules and may either be exclusive-use or shared-use. Only one unit at a time can use exclusive-use resources. Shared-use resources can be used by more than one unit at a time. If two units need an exclusive-use resource, one of them is going to be waiting a while.

If two units need a shared-use resource, they may both be happy. However, just because a shared-use resource is available to more than one unit does not mean it can accommodate all units at all times. For example, a glycol cooling system can be a shared-use system. However, the cooling capacity of such a system has its limitations. Two units requiring a total thermal transfer that exceeds the capacity of the glycol system might cause a slowdown in both batch processing times or, worse, an overload of the cooling system.

You must also be careful about how you design the use of these shared resources. Let's say unit A needs the glycol system and starts a pump. Then unit B needs the system. The pump is already running, but that's okay. When unit A is finished, it should not stop the pump or unit B will be affected. In this example, unit B should be responsible for shutting down the pump. This type of logic can easily be handled in equipment or control modules.

Arbitration is necessary when more than one unit or resource needs the services of another resource. This resource contention must be dealt with somehow. Here are some methods for doing so:

- Wait until the resource owner is finished with the resource. (First come-first served approach.)
- Attempt to find a substitute resource. This may include attempting to alter the path.
- Preempt the resource owner and acquire the resource based on a set of priorities. This can get complicated, especially if the resource needed contains material that is incompatible with the recipe requiring it.

All resource allocations can occur at the start of a recipe, effectively reserving resources, even though another recipe may require a reserved resource first. Allocation does not have to occur before a recipe begins. Dynamic allocation can occur as the batch is running. If more than one type of resource is available at a particular step in the process, a selection algorithm can be used to choose the best resource. Perhaps the material that makes up the resource (e.g., stainless steel versus titanium) will work with all recipes, but one material is preferred for the particular recipe running. Or one resource type processes quicker than another does. Allocation can be handled implicitly by S88 batch management software or explicitly by phase logic.

We've discussed equipment, recipes, and important control issues. Now it's time to begin exploring how to specify and design a batch management system. To do that we must first discuss what information is critical to a batch management system. That's coming up next in Chapter 9.

Batch Activities and Information Management (The Cactus Model)

Now that we've introduced the physical model, recipes, and how to link them, we think it's time to start talking about defining and designing a batch management system. That's the purpose of the next three chapters. Believe it or not, in our consideration of batch processing systems we've really been hanging out at a somewhat detailed level. Fasten your seatbelts, we're about to rapidly ascend to 20,000 feet. (That's about 6,096 meters for our readers using the metric system . . .)

This chapter introduces the control *functions* associated with batch manufacturing. "Wait!" you may exclaim, "haven't the last 87 or so chapters dealt with control functions?" Well, technically, no. According to S88, up to this point we've discussed control *tasks*. It appears as if the standard considers control *functions* to be a superset of control *tasks*. Additionally, control functions are grouped into control *activities*. So for review: a control activity is made up of control functions, which elaborate on control tasks.

We personally believe our readers would find it much easier if we referred to these three items as *batch management activities, batch management functions,* and *control tasks*. However, too many S88 committee members know where we live, so we're going to stick to the terminology in the standard.

The Control Activity Model

Figure 9.1 shows the control activity model. (If you listen very closely, you may hear committee members refer to this as the *cactus model*.) While it may not seem like much now, the control activity model provides an overall perspective on batch control and shows the primary relationships between the control activities.

One important aspect of the cactus model is that it provides a description of the information that is shared between the control activities as well as between the control functions within each control activity. Depending on your perspective, the type of information needed at various parts of your process can really help drive your design. Jim has a philosophy (and Larry is tired of hearing it over and over again) that if people are a company's most important asset, then information

Figure 9.1 The Control Activity Model

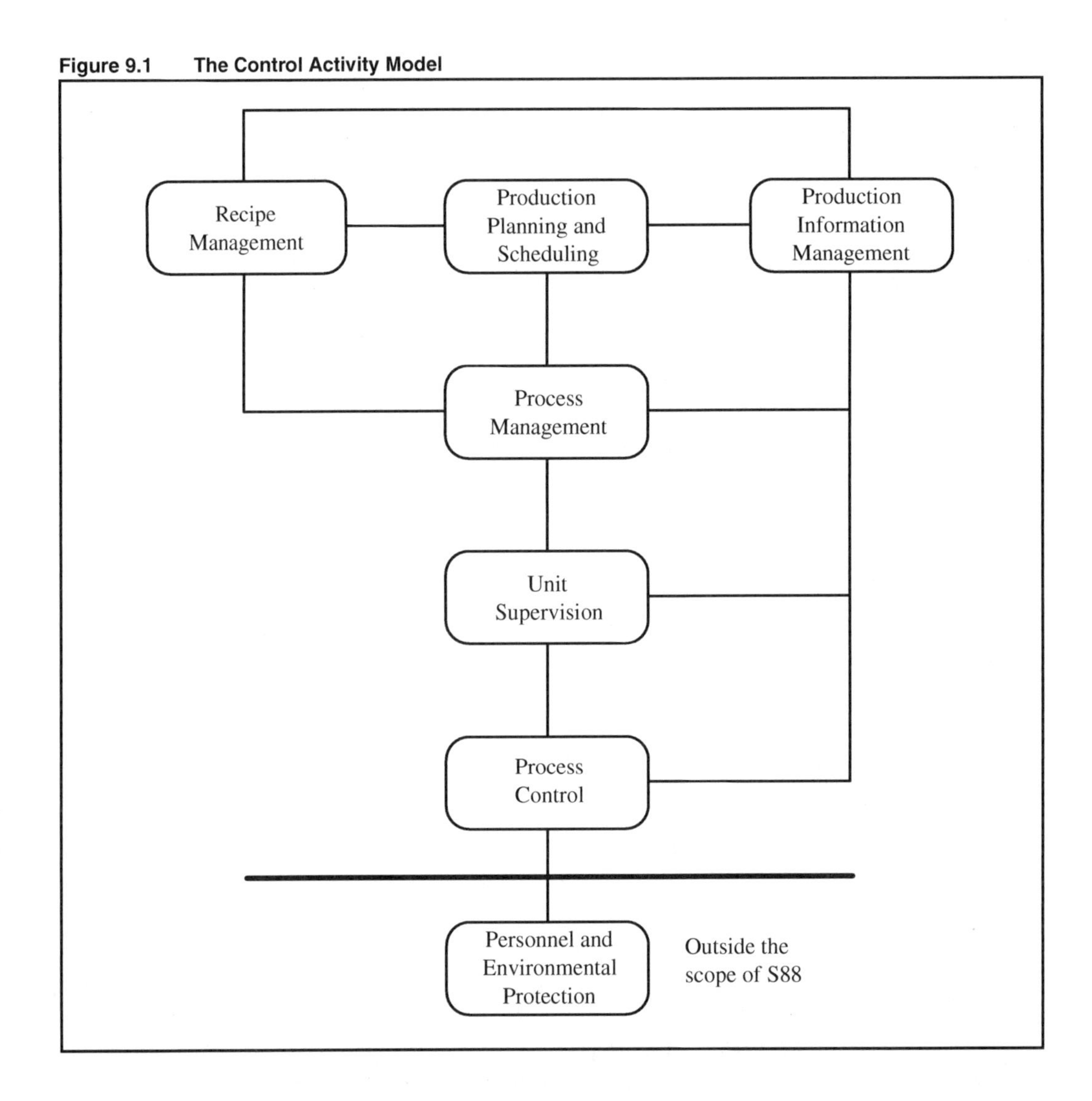

should be its second most important asset. However, in reality, information can be a company's number one quality problem when it's inaccurate, not timely, or not assembled correctly. If your manufacturing process is not defined well enough, you may not be collecting the right information you need to make good decisions. Get input from the right people on what information is needed. Don't guess; go to the source.

Remember that in Chapter 1 we said S88 can be applied regardless of the degree of automation? That rule also applies to elements of the cactus model. The shared information we're going to discuss in this chapter apply regardless of whether manual or automated systems are used to gather, store, analyze, and report the information. Figure 9.2 shows the cactus model with aggregate information flows that have been inserted between the control activities.

Figure 9.2 The Control Activity Model with Information Flows

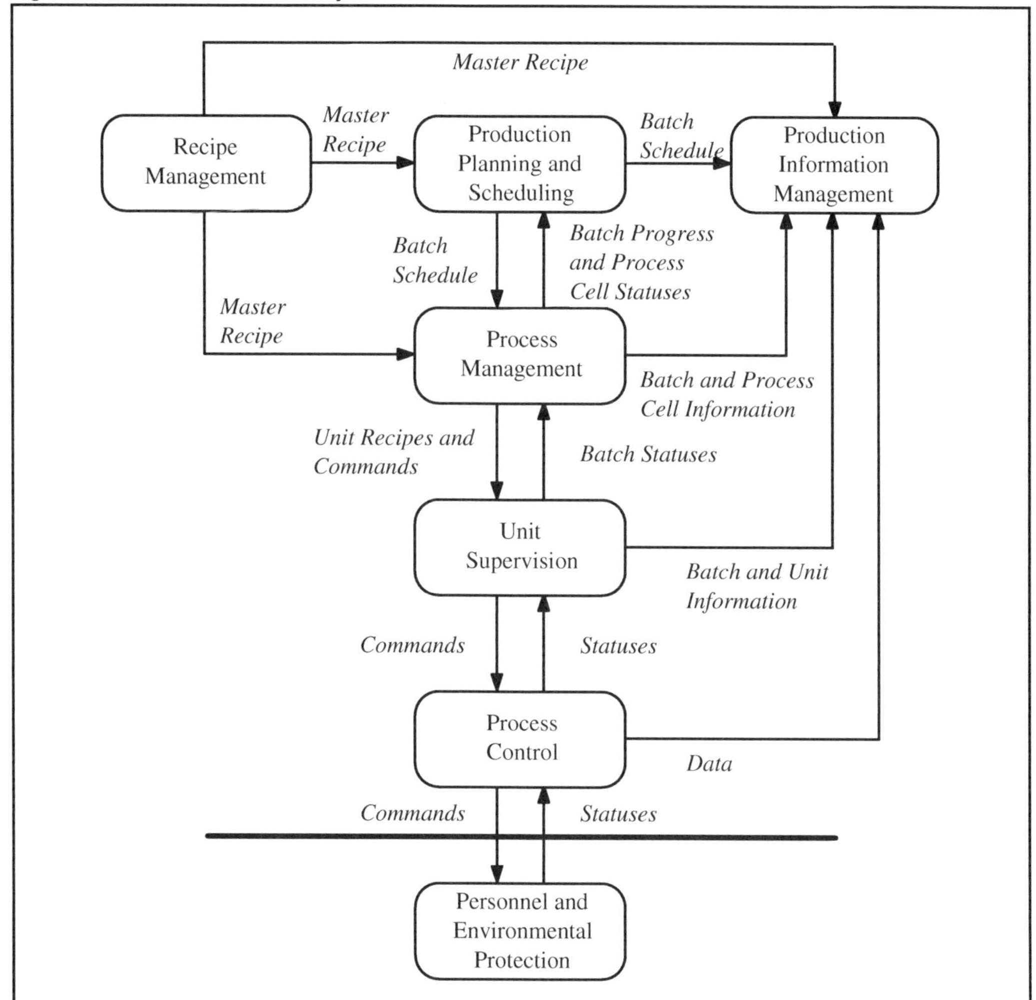

As we discussed earlier, the S88 standard breaks down each of these control activities into control functions. In the following sections we're going to provide an overview of these activities and appropriate functions. However, because batch management packages handle many of these functions and activities for you, we're going to let you read the standard for the details. (If we were college professors writing this book, we would say, "Understanding the associated detail is left as an exercise for the student.")

Recipe Management

The *recipe management* control activity contains control functions that create, store, and maintain general, site, and master recipes. The "deliverable" of this activity is a master recipe. The *process management* control activity uses the master recipe to create control recipes. (Recall that a process cell uses a master recipe. Control recipes are created by copying the master recipe and plugging in specific parameters that correspond to a unique batch.) Figure 9.3 shows the control functions associated with the recipe management control activity.

Figure 9.3 The Recipe Management Control Activity

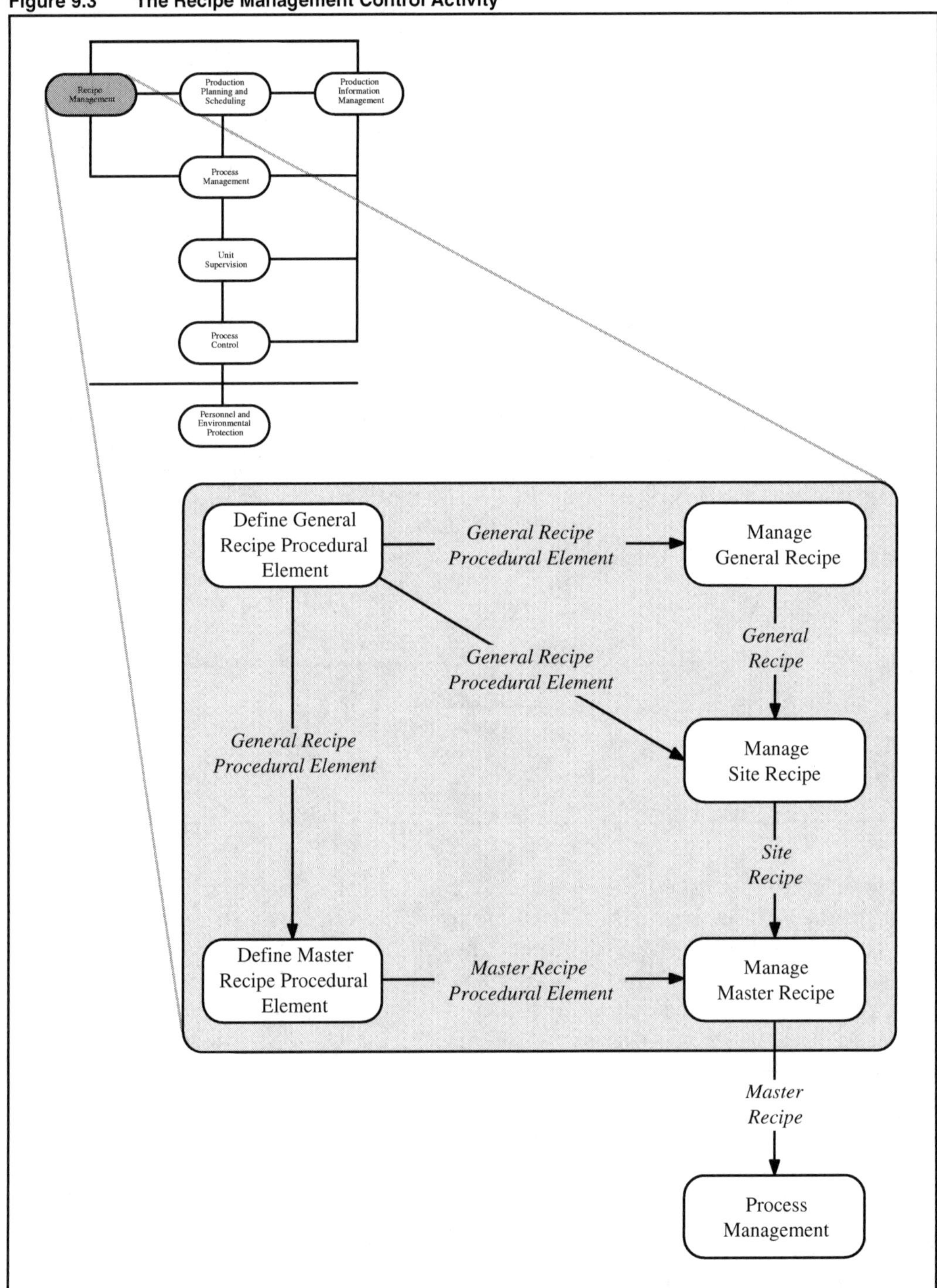

The *Define General Recipe Procedural Element* control function creates and maintains procedural elements that are used as building blocks in general and site recipe procedures. That is, this function takes care of the elements from the process model that define how general and site recipes are to be executed. (Can't remember those elements? Here's a clue: Process, Process Stage, Process Operation, and Process Action. See Figure 5.3.) These procedural elements are passed on to the *Manage General Recipe, Manage Site Recipe,* and *Define Master Recipe Procedural Element* control functions. (The procedural elements are made available to the *Define Master Recipe Procedural Element* control function so that the process intent of the general recipe is available for use in the master recipe.)

The *Define Master Recipe Procedural Element* control function is very similar to its general recipe counterpart, except that the procedural elements from this function are based on the procedural control model (Procedure, Unit Procedure, Operation, and Phase). This function needs to define the relationship between processes and procedures, process stages and unit procedures, process operations and operations, and process actions and phases. Master recipe procedural elements must reference equipment procedural elements when the derived control recipe is executed. Most commonly, this is done at the phase level. That is, the recipe phase must map to a corresponding equipment phase. Refer back at Figure 7.5 to see what we mean.

The *Manage General Recipe, Manage Site Recipe,* and *Manage Master Recipe* control functions are responsible for creating, maintaining, and storing their respective recipe types. Each function handles the selecting and combining of procedural elements to create recipe procedures, incorporate formula information, specify equipment requirements, and manage recipe changes. Note that site-specific procedural elements do not exist, so the *Manage Site Recipe* control function gets its procedural elements from *Define General Recipe Procedural Element.*

Transforming a site recipe into a master recipe can be complex. (Remember that the site recipe is based on the process model and the master recipe is based on the procedural control model.) The master recipe procedure must match the intent of the site recipe procedure.

From a practical standpoint, isolating the management of procedural elements from the management of the recipe allows you to reuse modular procedural elements. Well-designed, frequently used process actions, process operations, process stages, and/or complete procedures can make recipe transformations at lower levels much easier to accomplish and recipes more consistent.

OpenBatch is designed to work with an area (with multiple process cells). Therefore, it manages master and control recipes. Existing batch management packages do not handle general or site recipes. You may need to manage those yourselves. There are packages on the market, such as Sequencia's enterprise recipe management products, that manage enterprise and site level functions for you, including general and site recipes.

Production Planning and Scheduling

This control activity essentially accepts data from a corporate or site scheduling activity, from master recipes, and from resource databases (raw material, equipment, people, utility availability, etc.) and uses an algorithm to produce a batch schedule. This schedule is gladly given to the process management control activity.

Within *production planning and scheduling*, the control function *Develop Batch Schedules* describes what product is to be manufactured, when, in what quantity, and using what equipment. This control function accepts inputs from several sources:

- *Production or master schedules*—These explain what products need to be made, when they are to be made, and how much of each is to be made. In today's environment, this information may be generated by a finite scheduling software package.
- *Master recipes*—The recipes describe what raw material and equipment is needed. The equipment requirements are key since several different recipes that are run simultaneously may need the same equipment.
- *Resource databases*—Company or site materials requirement planning (MRP), manufacturing resource planning (MRP II), or enterprise resource planning (ERP) software can provide raw material inventory information (including packaging materials) and sometimes equipment availability information. Maintenance management systems (MMS) can provide information on equipment availability. Operator or other labor resources may be tracked by one of the above systems or by another HR-based system.

With these inputs, *Develop Batch Schedules* runs a scheduling algorithm to produce a batch schedule. The batch schedule should at least provide a prioritized list of batches to be made and the equipment, raw materials, and personnel needed for each. Having a detailed schedule helps you plan, but often manufacturing just doesn't run like you wish it would. Raw material quality problems, equipment failures, and personnel issues can all throw a kink in your plans.

Some S88 products, including OpenBatch, help to provide some process flexibility using a technique called *late binding*. Let's say you have a recipe that requires several units in a process train, and during the second process stage you can choose between three identical reactors. With OpenBatch, you need not commit to using a specific reactor until the second process stage is ready to execute. That way, if you planned to use reactor R1 but it had a temperature transmitter failure, you can use reactor R2 or R3 as an alternate without canceling your recipe in progress.

As with recipe management, production planning and scheduling is made up of several control functions. However, except for *Develop Batch Schedules*, they're all outside the scope of S88. Don't underestimate the importance of production planning and scheduling, however, because the capabilities of this activity can significantly influence plant throughput and equipment utilization.

Production Information Management

This is a heavy-duty control activity. It's the bubba that performs the collection, storing, processing, and reporting of batch production information. There are several control functions within the *production information management* activity, but the one that really counts is *Manage Batch History.*

The batch history is data that reflects the history of the batch. (Pretty straightforward, isn't it?) Depending on your batch package, the batch history may be organized into a single flat text file, into several files, or inserted into a database. As a user, you may have the option of choosing the disposition of the batch data.

The batch history is made up of entries. Each entry is a portion of information about the batch and represents one value or a set of values describing one batch event. (Don't get hung up on the formal definition of a batch event. Operator commands, the starting or ending of procedural elements, and exceptions are all examples of batch events.)

All batch management packages implement some form of data collection and reporting. OpenBatch creates one flat file per batch that it calls an "event journal." Sequencia also includes a report generation program that can format a pretty report from the event journal. The user also has the option of configuring an ODBC (Open Database Connectivity) database (such as Oracle, Microsoft SQL Server, or Microsoft Access) that can accept the contents of each event journal. This is convenient for running historical batch reports or performing some form of analysis on multiple batches.

S88.01 suggests the types of data to be collected with each batch event, as well as the data to be collected at the start and end of every batch. S88.00.02 goes one big leap further by *defining* a batch history information exchange format. Bravo! When we installed OpenBatch, it used a single database table to store all batch events. This is fine until data from your ten-thousandth batch is inserted into that table. (Our batches at Ben & Jerry's were relatively straightforward, and we amassed over four hundred events per batch! That's a lot of database table records!)

Up until S88.00.02, there has been no suggested database architecture for batch manufacturing. Making the valid assumption that every company (and perhaps even every different *site* within a company) will have different data structures for storing batch information, Sequencia chose to create a single batch history table to serve as an "interface" to existing data management systems (both computerized or paper).

Here are the database fields (columns) in the OpenBatch batch history table:

Table 9-1. OpenBatch Batch History Table

Local time	Time stamp for event
Batch ID	Operator-supplied batch identifier (such as lot number)
Unique ID	Batch ID supplied by OpenBatch that is unique for all batches for all time
Area	The area where the batch was made (this is the area from the physical model)
Process Cell	The process cell where the batch was made
Unit	The unit in which this event occurred
User	The operator who is overseeing the batch
Phase	The phase running that triggered the event
Event	The event name
Description	An additional description of the event (this varies with the event type itself)

S88.02 defines two database tables that provide a logical organization of the batch history data to be exchanged between systems:

- *S88_HistoryElement*—Contains data that might be common for many batches.
- *S88_HistoryLog*—Contains vital information for each batch. Some records might contain data unique to a single batch. Other records may reference data in *S88_History Element*.

By having *S88_HistoryLog* reference elements in *S88_HistoryElement*, instead of repeating the data in *S88_HistoryLog* the combined size of both tables can be much smaller than a single batch history table. (Note that this architecture is designed to work well with any of the relational database engines available. For those of you who are not database propeller-heads, it was the need for vast amounts of data that drove the development of relational databases. Instead of repeating common data, a separate table would contain often-used data. For example, a sales order system keeps track of all orders placed by many customers. Instead of repeating a customer's address for each separate order, a simple reference—a *relation*—would link each order to a single record in a separate table containing the customer's address. This saves an incredible amount of storage space.)

The same concept is used to reduce space in the batch history database. For example, let's say you wish to record the time when an agitator starts during a batch. Let's assume that the agitator starts at the same time or under the same conditions for every batch. Instead of placing that "start" record into a single batch history table for every batch, you can place it *once* in the *S88_HistoryElement* table and reference it by relation as many times as you need to in the *S88_HistoryLog* table. Over the long run, you save a lot of table space. Space may not be a huge concern, as disk drives are darn cheap nowadays, but database performance drops way off as tables grow in size. A batch history database based on a single table can grow very large very quickly. In our batch system at Ben & Jerry's, our single batch history table grew to over four hundred thousand records in just six months.

Regardless of the method used to store the batch history, some of the different types of events recorded include:

- Name of recipe, author, and revision
- Name of physical model (area model) and revision
- Operator commands issued
- Procedural element starts/stops
- Formula parameters (for example, targeted ingredient quantities to add)
- Exceptions (failures)
- Phase-generated messages and reports (for example, actual ingredient quantities added)

The last event type—phase-generated messages and reports—is very important. These customer-defined events gather and report the specific information you need for your process. At the ice cream plant, we chose to record items like the actual ingredient quantities added, the tanks used to source ingredients, and when a batch tank agitator started.

Those of you presently using other data-gathering tools, such as data historians, may wish to give serious thought about which data should be gathered by a batch management package and which by the data historian. Some of it may come down to money questions—which tags are more expensive? You may also wish to consider convenience. For example, recording the temperature profile for a reactor may be better suited to a data historian than to a batch management package. However, all of this completely depends on your needs.

Process Management

The *process management* control activity is a collection of control functions that manages all batches and resources within a process cell and collects batch and process cell information. These control functions are shown in Figure 9.4.

Figure 9.4 The Process Management Control Activity

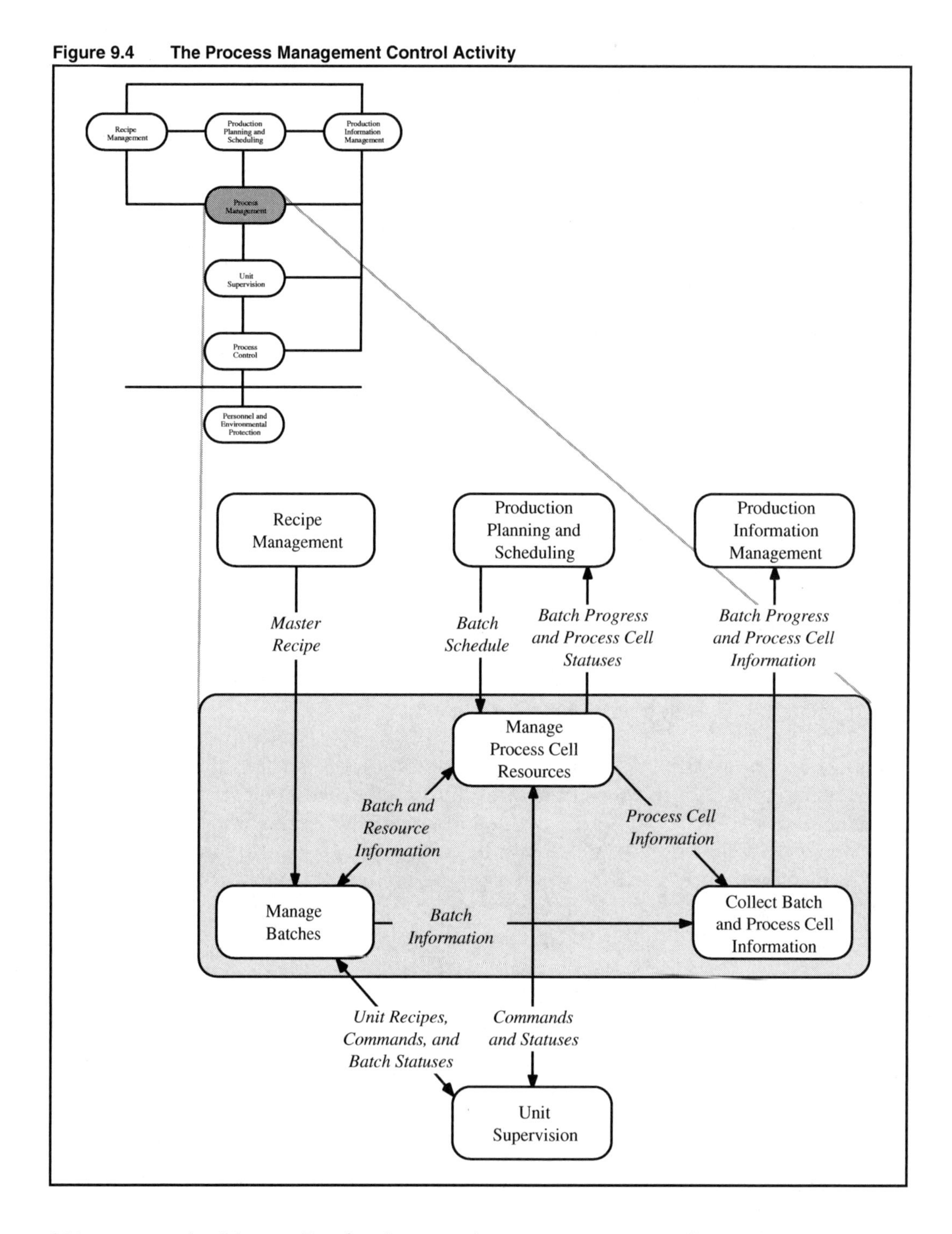

S88 requires the *Manage Batches* function because a process cell may have multiple batches running in multiple units. This function creates a control recipe from a copy of the corresponding master recipe. It assigns a unique batch ID to the batch and control recipe and performs scaling to size the control recipe to meet the batch quantity desired. *Manage Batches* requests and releases units and other equipment as the recipe requires. It processes requests for mode or state changes (such as putting the batch in hold or running a batch in manual). Finally, it maintains batch status information for use by the *Collect Batch and Process Cell Information* control function.

The *Manage Process Cell Resources* control function allocates and reserves units and other equipment and arbitrates multiple requests for the same equipment. It also manages material (ingredients and completed product) within the process cell.

Collect Batch and Process Cell Information collects batch and equipment events from *Manage Batches* and *Manage Process Cell Resources* and sends that data to the *Production Information Management* activity.

Unit Supervision

This control activity, which links the recipe to equipment control, is shown in Figure 9.5.

Manage Unit Resources functions mainly to acquire and release equipment modules. *Acquire and Execute Procedural Elements* retrieves the unit recipe (the unit procedure portion of the control recipe) from *Process Management*. It then determines which procedural logic to run (unit procedures, operations, and/or phases), which parameters to use, and which equipment to use and gathers pertinent batch information (such as batch ID, name of product, and equipment constraints). *Collect Batch and Unit Information* gathers data about unit supervision and sends it to *Production Information Management*. The data collection may be conditional. That is, not all data may be collected at every given time interval or at every batch event.

Process Control

The *process control* activity rules over procedural and basic control (see Chapter 7 for a definition of these control types). It also does its share of collecting and forwarding data. The device manipulating the process (a PLC or DCS) can handle much of this control activity. See Figure 9.6.

Execute Equipment Phases sends commands and passes parameters to equipment entities (units and equipment modules). It is initialized by *Acquire and Execute Procedural Element* in the *Unit Supervision* activity (see Figure 9.5). *Execute Basic Control* commands changes in equipment and process states by sending commands to control modules, such as PID (proportional-integral-derivative) controllers or totalizers. Commands may come from equipment phases or from operator-initiated manual control. *Collect Data* retrieves data from sensors or from events that occur within the process control activity. As with the *Collect Batch and Unit Information* function of the *Unit Supervision* activity, data collection may be conditional.

Figure 9.5 The Unit Supervision Control Activity

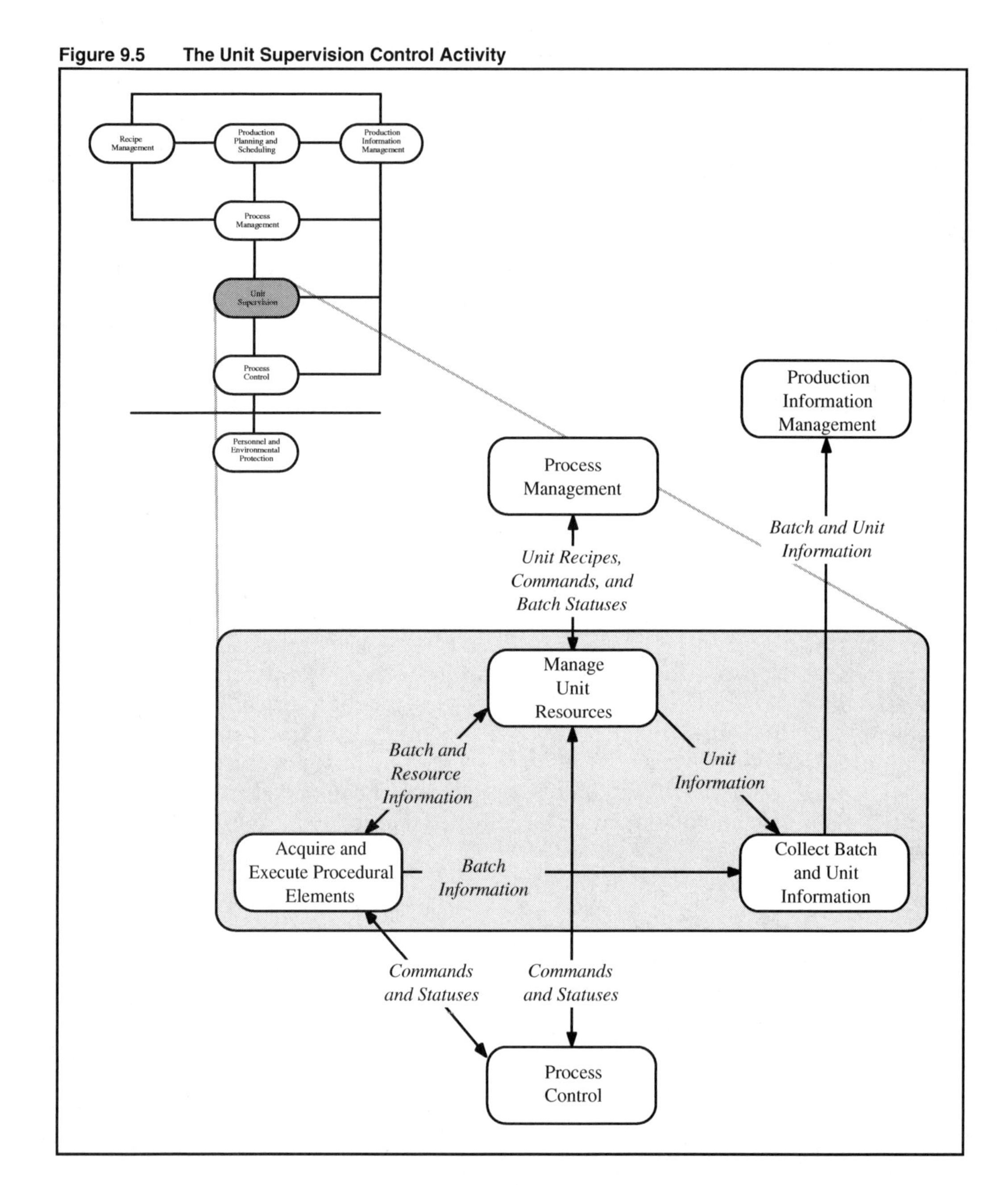

Figure 9.6 The Process Control Control Activity

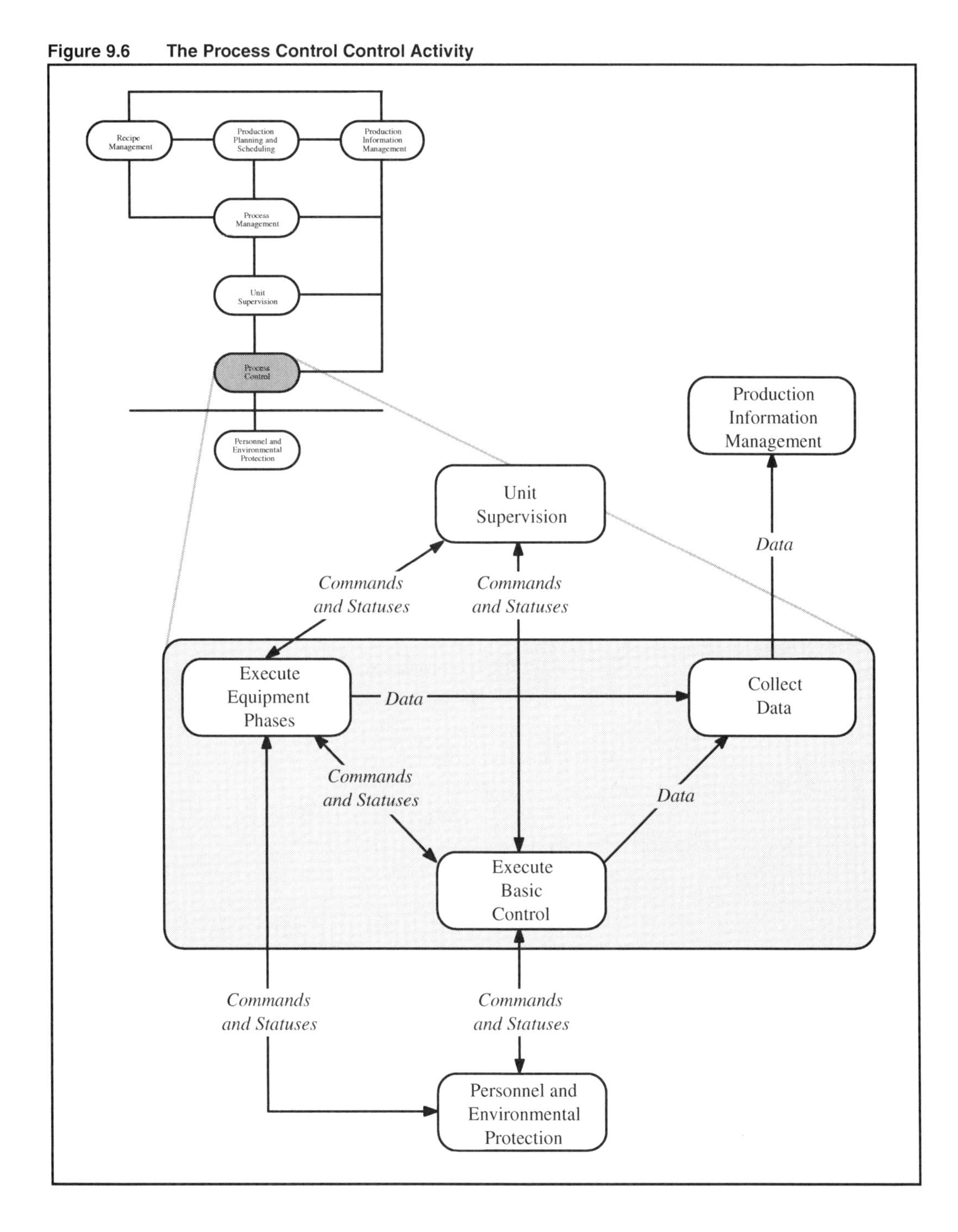

Personnel and Environmental Protection

This control activity is outside the scope of S88 but certainly not outside the scope of the batch control system. Operator safety should be every control engineer's number one priority (no kidding here). Environmental and product safety runs a close second. Emergency stops and safety interlocks are examples of what this activity covers. S88 references another ISA standard for more information: S84.01, *Application of Safety Instrumented Systems for the Process Industries.*

Presenting Information to the User

Batch systems typically present users with information in the form of human-machine interface (HMI) screens or reports. This book isn't about designing a graphical user interface, but one thing you might want to consider is installing a diagnostics screen on your HMI that indicates the current control step of each phase. (If you're interested in using the HMI tools included with batch management software, some packages—including OpenBatch— provide control step information on operator screens included with the package.) With good phase documentation (see the next chapter) these indicators will make great troubleshooting tools.

As we mentioned earlier, batch management packages provide varying levels of reporting capability. You may wish to view reports on a computer screen or print them. Just as OpenBatch downloads parameters into equipment phases in PLCs or DCSs, it can also upload *phase report values*. A phase report value is typically one specific item of information, such as the total gallons of cream actually pumped. Don't confuse printable batch reports with uploadable OpenBatch report values. These uploadable values trigger events that become entries in the event journal and maybe entries in a database batch history table. However, just like downloadable parameters, each phase report value eats up a tag. The overriding goal is to have an electronic batch record (EBR) that works well for you, but as users we also know the tug of war of functionality versus time and money. So, unless you have an unlimited budget, if you don't need the report value don't use it. In the immortal words of Sheriff Buford T. Justice, "You can think about it, but don't do it."

The type and style of your printable reports is up to you. You'll probably design a report that mimics a report already in use by your operators or management. Perhaps your batch management system will simply feed information to an existing reporting system.

Keep in mind that you may wish to have some type of data filter for your reports or for archiving to a reporting system. Most of the time, your operators or an existing reporting system will want to know about recipe parameters and phase reports (e.g., ingredients and amounts requested and received). The order of ingredients may be pertinent, and time stamps could be important. Knowing exactly when each phase started and stopped, however, may not be critical to your reporting requirements.

We created three reports when we installed OpenBatch at Ben & Jerry's:

- *Batch history*—This showed all data collected in the event journal, the whole enchilada. This came in handy for tracking down process issues.
- *Batch summary*—The essential, but minimal amount of data associated with a batch: recipe name, operator name, batch tank used, batch start and stop times, ingredient names, targeted quantities and actual quantities, as well as agitator speed.
- *Run summary*—This took data from each batch summary and combined it in a matrix to summarize an entire run of batches for one type of mix.

You probably have other reporting needs. Those of you in highly regulated industries may find it useful to create specific reports tailored for inspectors.

OpenBatch collected all the data for each batch. We used specific database SQL (structured query language) statements and VBA (Visual Basic for Applications) to filter and format the data for batch summaries and run summaries.

As we mentioned in the introduction to this book, we were able to get rid of our clipboards in mix-making by taking advantage of automatic data collection and reporting. Removing the operator from the manual handling of data generally removes a non-value-added task. Data is more accurate, timelier, and more available. This will allow you to make more money.

Before you get too scared about VBA or SQL, understand a couple very important points. First, learning VBA or SQL is not that difficult, but it will take some time. Second, there are many tools available that can help you without having to learn the details of VBA or SQL. Macro recording functions in all of the Microsoft Office products handle the creation of a lot of VBA code for you, and products like Microsoft Query, Business Objects, and Crystal Reports allow you to query databases without knowing the intricacies of SQL.

Okay, time to descend to 10,000 feet . . .

10

System Specification and Design (Some of It, Anyway ...)

Okay, we've discussed requirements gathering for S88, as well as financing, equipment, procedures, recipes, modes, states, data collection, and reports. Throw in a partridge in a pear tree and you probably can write a song. In this chapter, we want to talk about specifying and designing a system.

If you haven't already done so, you may wish to consider attending a course on applying S88. Check with ISA, your favorite S88 consultant, or your software provider for training. For the past several years, the World Batch Forum has also included an S88 tutorial session as part of its annual "Meeting of the Minds" conference.

Most of you will probably specify a system as part of a bid package to original equipment manufacturers (OEMs) or integrators. Then you and your chosen vendor may jointly design the system. Perhaps you will want to work with a consultant to design the system on a fixed-bid price and then send that design out in a separate bid package for fixed-bid installation quotes. We did that a few times with other projects.

Chapter 2 discussed gathering functional requirements from your customers. Your deliverable is a *functional requirements document*. Don't confuse that document with the *functional system specification* that you'll give to a vendor (or vendors). We like to think of the functional requirements document as an internal (sometimes confidential) document to the manufacturing company. It is the basis for the functional system specification that you give to vendors. However, the functional spec is generally a superset of the functional requirements and provides more technical and project detail to your vendors.

How the level of detail in your functional specification compares to the level of detail in your design depends on the payment plan you expect your vendor to accept. If you want a fixed-price plan, your functional specification had better be detailed enough to avoid confusion and serious contention during the project. It goes almost without saying that your project has a better chance for success if you choose an OEM, vendor, and/or consulting firm that truly understands S88 *and* knows your industry.

Creating a Control System Functional Specification

In your functional specification you will want to apply everything you know about S88 so far, especially what you've learned about the standard's models and terminology. Define your areas, process cells, units, equipment modules, and control modules. Use the terms *procedure, unit procedure, operation*, and *phase*. Talk about master and control recipes. Ensure that your potential vendors understand how important it is to follow the standard. Ideally, you'll choose an OEM or integrator that already understands S88 and has several successful implementations under the belt.

Here's a tip: insist upon modularity of procedures and equipment entities, so they can be used and cloned—not only in this project, but also in others to follow. Object-oriented design techniques can help in this type of work. (If you don't know what object-oriented design is all about, learn about it!)

The control activity model, as we discussed in Chapter 9, is a great basis for defining the control functionality you need. Use the information from Chapter 9 and additional detail from S88 to think about what you need to specify in your functional description. Remember these key control activities: recipe management, production planning and scheduling, production information management, process management, unit supervision, process control, and personnel and environmental protection.

If you're retrofitting an S88 solution into an existing system, your physical design may very well be unchangeable. If this is the case, make sure you have attached updated P&IDs (process and instrumentation diagrams), electrical drawings, and information system requirements in the bid package. If you hadn't already done so when you created your functional requirements document, take the time to find out *exactly* how your operators run the process. (No matter how smart you think you are, good operators will always know the process better than you will.) After you do that, find out exactly how the operators would *like* to run the process. These may be two entirely different descriptions.

For better or for worse, S88 is not a compliance document. That is, you do not have to follow every letter of the standard to say you are S88-aware. But neither do S88 software vendors. This is why there are such a wide variety of S88 implementations in industry. Use this to your advantage by balancing the concepts in S88 against what is practical and economically feasible for your operations.

The rest of this chapter discusses specific items that you can include in your functional specification.

Documenting Physical Equipment

Your batch control system will (or should be) limited to a single process cell. If it's not, you may wish to rethink the definition of your physical plant. Start segmenting equipment entities based on the physical model:

1. First, *find* your units.
2. Next, determine where your equipment modules and control modules are.
3. Finally, make sure each of your equipment entities performs a clear, independent task, and remember the rules for what defines a unit. It is very common to have similar units, equipment modules, or control modules that operate in parallel.

Specify the behavior of the equipment entities in accomplishing your batching tasks. That is, define the requisite basic, procedural, and coordination control for each entity. Remember the big difference between the way S88 defines equipment modules and control modules: equipment modules are allowed to execute phases; control modules are not. Chapter 4 discussed choosing and defining equipment modules and control modules, but here are some examples of equipment modules:

- A pump, one or more valves, and a flowmeter that comprise a recirculation loop
- Storage tanks, outlet valves, a common header, and a pump that feed an ingredient to one or more batch tanks
- A level transmitter and an agitator responsible for mixing ingredients in a batch tank

Finally, don't forget about important data that you need to collect or send to each of these entities. This information will help define your instrumentation and control requirements.

Documenting Master Recipes

Develop your master recipes using S88 definitions: header, equipment requirements, procedure, formula, and other information. As with batch histories, the forthcoming S88.00.02 standard will provide great detail about developing the data structure for recipes, including using a database to store recipes. Once S88.00.02 is published (which is expected early in 2000), consider leveraging it in your design. However, if you are leaning toward a particular S88 commercial software package, it may be much easier for you to specify recipes based on the capabilities of that package.

Sequential function charts (SFCs) work well for documenting recipe unit procedures and operations. Figure 10.1 is a recipe operation, first shown in Chapter 5 (see Figure 5.8), that is described using an SFC.

Figure 10.1 A Recipe Operation Documented Using a Sequential Function Chart

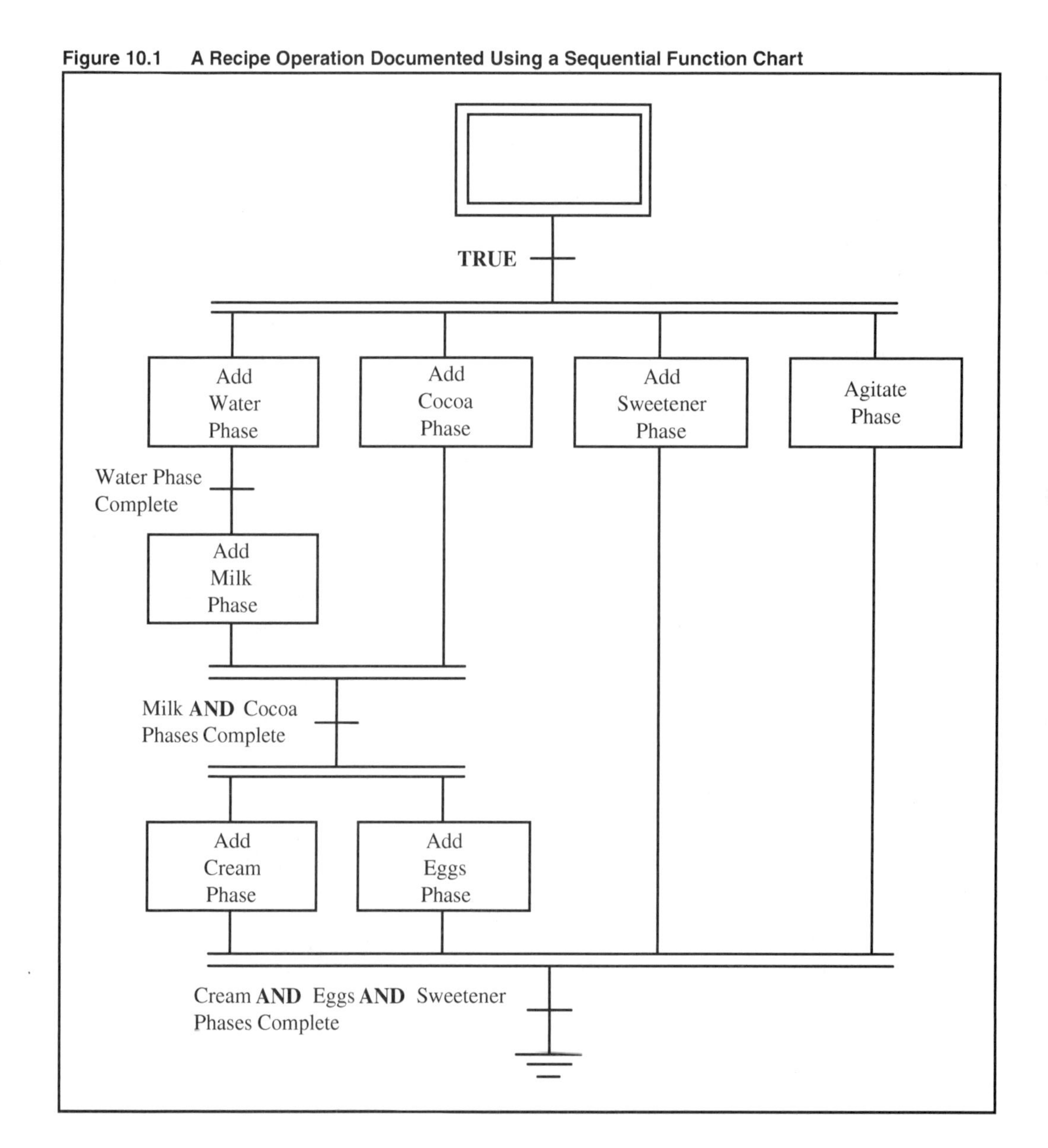

If your product moves in a straightforward fashion from unit to unit, SFCs are also a good way to document recipe procedures. But if there is a lot of unit interaction, such as coordinated material transfer from one unit to another, you may wish to consider a more time-based documentation method. Some companies have used Gantt-type charts to document recipe procedures (these are similar to Gantt charts used in project management tools). Figure 10.2 shows such a chart.

Figure 10.2 Recipe Procedure Documented Using a Gantt-type Chart

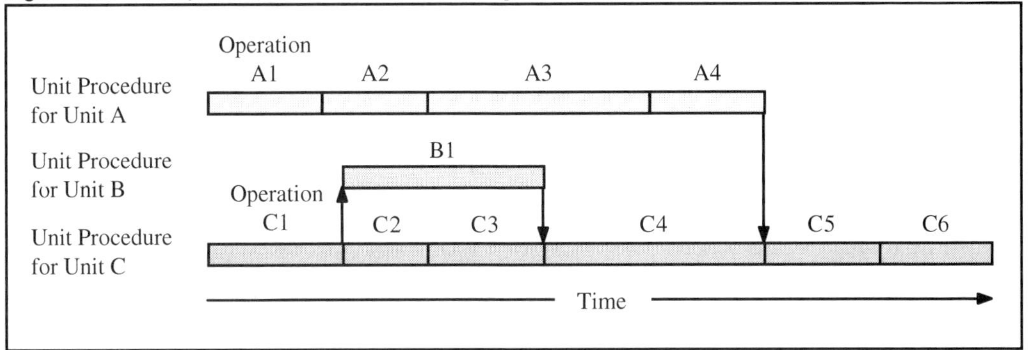

Documenting Equipment Control

As we discussed in Chapter 7, most equipment control will consist of *equipment phases* that are linked to *recipe phases*. Properly describing equipment phases is very important because you're describing how the equipment will physically run. Also realize that at least 60 percent of the content (and effort) of your functional specification will involve defining equipment control. Because this percentage is so high, we've decided to create a separate chapter on equipment phases. So let's finish up this chapter and then move on to Chapter 11.

Documenting Other Important Aspects of Your Batch System

Don't forget about important issues like your batch schedule and batch history. Decide if batch scheduling is going be handled manually or by a software package. If it's handled by another software package, how will the output from that software be fed into your batch control system, if at all? For your batch history, decide what data you wish to collect and report. From that, determine the format of the information and what information should be filtered and sent to higher-level systems. Above all else, make your reports look pretty. *Important announcement*: this brings us to Jim & Larry's general order #1—Image is everything. A well-formatted report (or a snazzy presentation) will *always* add value to what you're suggesting, selling, or recommending. In those inevitable times when you are rushed, a good-looking report or presentation can often distract your audience from the fact that you may not have every one of your ducks in a row.

A Sensitive Subject: Working with Your IS/IT Department

We would be remiss if we didn't discuss a subject that is probably very sensitive in your organization: the joint efforts of engineering and information systems (or information technology) departments. For some reason, engineering and IS/IT always seem to fight like cats and dogs. We're not philosophers, nor are we psychologists, so we're not going to try to understand why. We've lived through those experiences ourselves. Sometimes it just isn't a pretty sight.

Anyway, bury the hatchet if possible because your IT department may have the skills and expertise that you need to deal with databases, reports, computer hardware, networks, and server operating system administration. If your organization is really progressive and you have some form of production execution system (such as an MES software package), the IT folks may very well own that system. Your system may need to talk to their system and vice versa. And consider this, too: from the IT point of view, a batch management system might fall within the "information domain"; at the very least, it crosses the boundary between the control and information domains. More than likely, you will need the support of your IT colleagues to make your system a success.

We've often found that basic problems arise because of misunderstandings over terms, like *process control*, *batch*, and *real time*. (Try asking your IT neighbor what a batch is.) The more engineers and IT professionals can just sit down and talk, the better off both groups will be. Just as S88 has helped define common terminology for batch manufacturing, another ISA committee, SP95, is defining models and terminology for the purpose of integrating control systems with information systems.

ONE FINAL NOTE

And this is a biggie. *Host periodic reviews of your design.*

If you're designing the system internally, meet with peers from other departments or other sites. If you've handed over a lot of the design responsibility to a vendor or OEM, meet with that company often. There's nothing quite like the angst you feel at the end of the design phase when you realize the system cannot possibly do what you need it to do. Keep on top of things by staying involved and seeking help from others.

11

Specifying and Designing Equipment Phases

It's worth noting that this chapter has more text than any other. As entertaining and informative as we try to be, we are engineers, not motivational speakers. So, you may wish to grab your favorite caffeinated beverage before proceeding …

First, let's manipulate the space-time continuum and experience a temporal loop with this statement: properly describing equipment phases is very important because you're describing how the equipment will physically run. Also, as we said in the last chapter, realize that at least 60 percent of your functional specification content (and effort) will involve defining equipment control.

Within your equipment phases, you must consider important items like:

- *Phase logic*—The guts of the equipment phase.
- *Modes* and *states*—For example, how do you want the equipment to behave if it is transitioning from a *Run* state to a *Hold* state?
- *Allocation* and *arbitration*—These are necessary to properly coordinate the use of equipment modules.
- *Unit-to-unit synchronization*—For example, if performing a function such as transferring material from one unit to another, both units have to be ready before the transfer can begin.
- *Exception handling*—What will your equipment do if a flowmeter registers no flow, or if feedback says that a valve did not open?
- *Data collection*—What items do you want to report back to the event journal or database?

We will discuss each of these items in this chapter.

A Phase Review

First, a flashback review session:

- The instructions in a control recipe procedure must somehow be conveyed to the control system that is running the equipment. (Remember that a control recipe is the element of the recipe model that is specific to a particular batch.) Control recipes and equipment control can both have procedures, unit procedures, operations, and phases (see Figure 7.3).

- A link must exist somewhere that ties a control recipe to equipment control. The most popular point for this link is at the phase level, where a reference is made to an equipment phase that is executed to perform a particular batching function (see Figure 7.5). Remember that equipment phases are generally run in PLCs or in a DCS, but they can also run on a personal computer equipped with such tools as PLC emulation software. OpenBatch even has the capability to execute phases written in Visual Basic on a personal computer.
- Linking at the phase level really means the batch management system contains *recipe* procedures, *recipe* unit procedures, *recipe* operations, and *recipe* phases, and the equipment control system contains *equipment* phases.
- These equipment phases execute equipment control by running phase logic.

As part of your physical design, you identified your units, equipment modules, and control modules. As part of your recipe design, you identified your recipe procedures, unit procedures, operations, and phases. Since most batch management packages link recipes to equipment control at the phase level, phases are the key to physically running equipment. To put it very simply—and we could provoke argument from S88 committee members for saying this—operations and unit procedures exist so that you can group and sequence the phases necessary to make batches.

Recall from Chapter 5 that there are some rules for using phases:

- A phase is the smallest element of procedural control that can accomplish process-oriented, yet product-independent, tasks. In making mix at Ben & Jerry's, our phases perform basic and independent actions like transferring sweetener and agitating a batch tank.
- A phase can operate on more than one set of equipment, such as a unit, but not at the same time.
- More than one phase can control a piece of equipment.
- A phase may require that one or more other phases be running to perform its task.

Defining Phases

Identifying the devices that will be controlled by phase logic is not difficult. Grab the P&IDs for your process cell and highlight the equipment associated with the process action you wish to accomplish. For example, for our *CHARGE_SWEETENER* phase, we highlighted the path from the sweetener storage tank to our batch tanks. In this path were the sweetener storage tanks, level probes for the storage tanks, valves, pumps, flowmeters, the batch tanks, and level probes for the batch tanks. (Let's not forget the pipe.)

Designing Phase Logic

Phase logic often breaks down into a series of *control steps*, with each control step possibly broken down further into *control actions*. If you talk to ten different people about control steps and actions, you may be lucky enough to get ten different viewpoints on the scope of responsibility for steps and actions.

We think control actions should perform simple, independent stuff, like opening a valve, specifying a set point, or performing a basic calculation. We think of control steps as logical groupings of control actions. One way of writing PLC phase logic is to consider each ladder rung as a control step and every "coil" branch as a control action. Rungs may tend to get long if several control actions are part of a control step, but this rung approach can be a convenient way to write code.

An easy way to describe a phase is to use a table. Table 11.1 describes a slightly fictitious *CHARGE_INGREDIENT* phase.

Table 11-1. A Simple Design for the CHARGE_INGREDIENT Phase

Control Step	Control Action	Comments
Initialize Phase	Reset reporting registers.	Reset registers for the start of new phase instance.
	Clear registers and flags.	Reset registers and flags used in phase logic.
	Download parameters.	Retrieve source tank and target quantity from batch management system.
Prepare Equipment	Reset totalizer.	Clear flowmeter totalizer.
	Adjust target quantity desired based on process spill factor.	Spill is difference between actual and targeted quantities when transfer is complete.
	Put new target quantity value into report to be uploaded.	It may be pertinent to store the adjusted target quantity.
Set Path	Open tank outlet valve.	
	Open header valve.	
	Open batch tank inlet valve.	
Transfer Ingredient	Start transfer pump.	
Transfer Complete	When flowmeter totalizer value equals target quantity parameter, stop transfer pump.	Delay closing valves to allow pump to spin down.
	Close all valves.	
Housekeep	Calculate spill for next transfer.	Spill is amount of extra ingredient obtained due to variation in process (pump wind down, slow-closing valves, etc.).
	Report actual amount of ingredient transferred.	

Don't Cry Over Spilled Mix

Sidebar time: Table 11.1 describes a condition we call spill. To us, spill is extra ingredient received because of variability in the process. When a totalizer hits its set point, valves don't close instantaneously, and pumps don't generally stop within a few milliseconds. When designing our system at Ben & Jerry's, we wanted the operators to get the quantities they asked for. (We did not want to burden the operators by forcing them to adjust their target quantities to get the actuals they needed.)

We calculated spill by subtracting the target quantity to be transferred from the actual quantity transferred. If a few extra gallons are fed into the batch tank, that number will be recorded in the spill register for that ingredient. The next time that ingredient transfer phase runs, that calculated spill will be subtracted from the target quantity to create an adjusted quantity that will be fed into the totalizer set point. If the process is not too horribly inconsistent, the actual quantity during this next phase run will be closer to target.

That calculation of spill worked for us, but we must admit that our calculation is rather simple. We know some batching operations that take a percentage of spill (80-95 percent) and use that to adjust the target quantity. This provides perhaps a smoother (asymptotic) tracking for getting the actual quantity needed. If you want to get sexy, you probably can go as far as creating some funky PID algorithm to calculate spill. And now, back to our discussion of equipment phases.

Describing equipment phases in tables works well in some cases, especially when asking vendors to bid on a design or design/build project. However, Table 11.1 is simple in that it does not address any exception conditions. You may want to consider what happens if and when the following occurs:

- The flowmeter registers no flow. (Did the pump start? Did the valves all open? Is there an open pipe connection somewhere? Is the flowmeter busted?)
- The ingredient tank empties before the target quantity is reached. (If you have more than one storage tank per ingredient, do you wish to identify a secondary tank to automatically pull from when the primary tank empties?)

Another great way to document phases is with a sequential function chart (SFC). You can use tools like Visio or AutoCAD to capture the SFCs. As Figure 11.1 shows, each SFC step is a control step and is numbered. The numbering of control steps works very nicely for PLCs since they can store and manipulate numbers much better than alphanumeric characters. Numbering the control steps by tens was our preference. You can number them any way you wish. However, unless you truly are a genius and never make any mistakes or you like renumbering control steps, you may wish to leave spare numbers between steps during your initial design. (Anyone remember classic BASIC with line numbers? It's like that but with no renumber command.)

The function of each control step is described to the right of the control step box. (That is, the control actions are described to the right of the box.) The transition condition necessary to execute each control step is described to the left of the transition.

The SFC way of documenting provides a couple of neat features that may be more difficult to describe in a simple table, like the one shown in Table 11.1. First, you can identify the condition or conditions necessary to transition from one control step to the next. (Now, we know you can add these two features to a table, but it starts to get messy.) Second, you can show the procedural flow within the phase. Figure 11.1 is a sequential function chart for what many call the *Running* logic—otherwise known as the default or normal logic—for the *CHARGE_INGREDIENT* phase.

Notice that Figure 11.1 handles a no-flow condition. Step 40 starts the transfer pump and also resets a delay timer. At Ben & Jerry's, we often used delay timers of somewhere around 30 seconds, giving the process that amount of time to get flow going before triggering a flow failure. If no flow was registered after that delay, we set a failure flag.

OpenBatch provides a way to enumerate failures for phases and requires a reserved register for each phase to hold a failure code. In our case, when we detected a flow failure we set the PLC phase failure register to 2. OpenBatch monitors each of the failure registers. When it detects a nonzero, it automatically places the phase in *Hold* and informs the operator that a failure has occurred.

Setting up a primary/secondary ingredient tank arrangement is not difficult. Somehow the equipment phase must be made aware of the chosen primary and secondary tanks. This can be handled with parameter downloads, or the operator can directly place identifiers to the primary and secondary tanks into PLC or DCS memory using an HMI. Here are some things to consider when using primary and secondary tanks:

- You will need to differentiate between flow paths for different tanks. You may have a common header and transfer line, but each tank probably will have its own outlet valve.
- You may wish to record the amount of ingredient transferred from each tank rather than just the total. You may also wish to record exactly which tank was used and when.
- Switch from the primary to the secondary tank when you register a no-flow condition (and you have given your process plenty of time to start pumping), when your tank instrumentation indicates the tank is empty, and when you have not transferred the target amount of ingredient.

Figure 11.1 Sequential Function Chart for the *CHARGE_INGREDIENT* Phase *Running* Logic

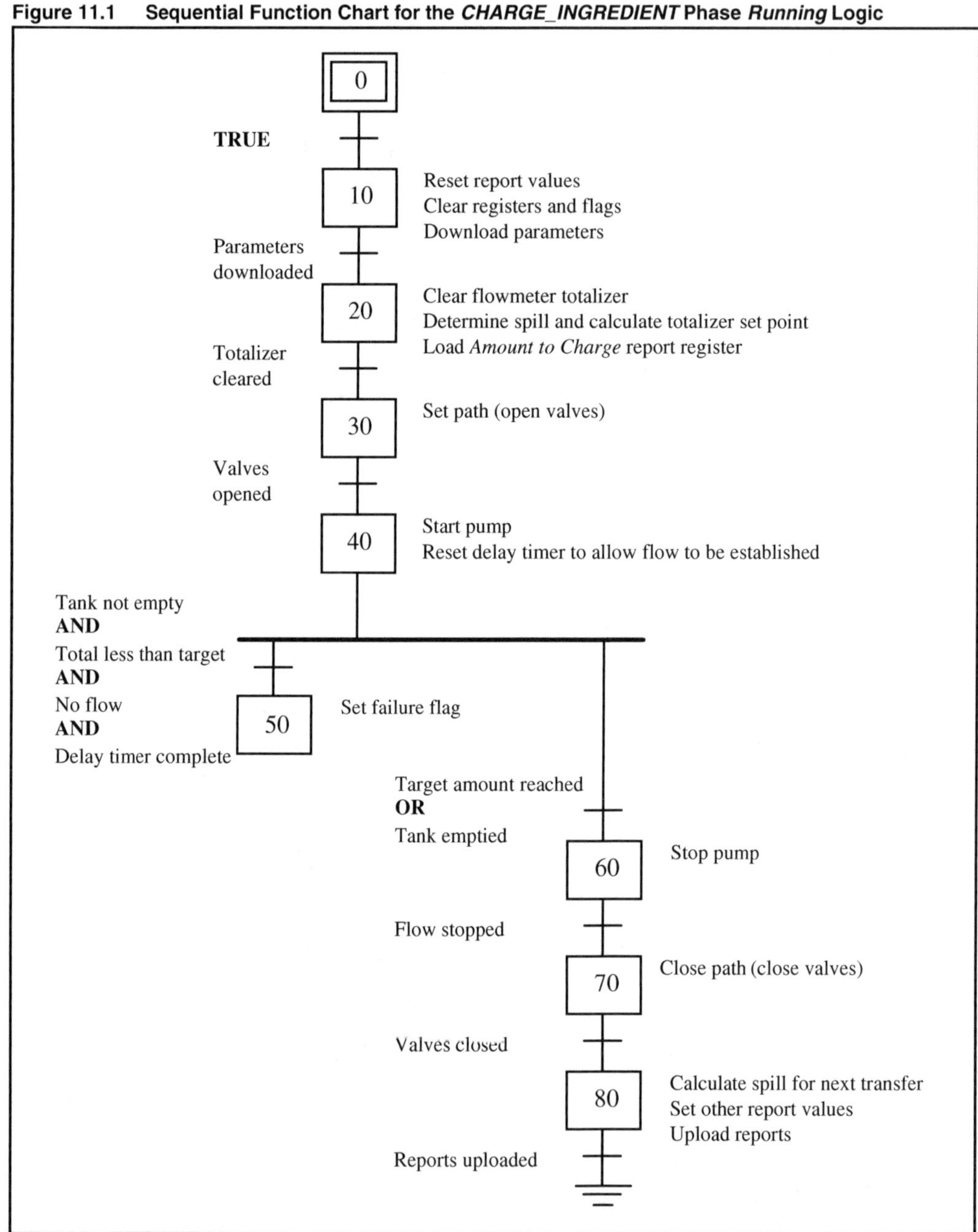

(Careful: a no-flow condition also could be caused for other reasons, such as a broken connection. Watch how you write your code.)

- Be prepared to prompt the operator for a secondary tank if he or she forgot to enter it into the HMI or into the recipe. (You may be able to verify that a secondary tank has been identified before allowing a batch to start. We couldn't at Ben & Jerry's because sometimes we just didn't have—or need—a secondary tank for each batch.)

- You'll need enough scratch memory to recalculate a new target quantity, especially if each tank has a different flow totalizer.
- You may not wish to calculate spill in the case of a tank switch. Just carry over the spill used from the previous complete transfer.

Modes and States

If you think we've exhausted our discussion on phase logic, think again. Phase logic is broken down into five different sections, one for each of five transient states: *Running*, *Holding*, *Restarting*, *Stopping*, and *Aborting*. Up until now, we've really only been talking about the *Running* state phase logic. If all goes well when you make your batches, you will only need to execute the *Running* state logic. However, that will not always be the case. For example, if an operator needs to hold a phase or if an exception occurs that needs to hold a phase, the *Holding* logic and probably the *Restarting* logic will run.

If you're saying to yourself, "Uh, what are these guys talking about?" re-read the section of Chapter 8 titled "States and Commands." It describes each of the "-ing" states in much more detail. If you're saying to yourself, "Uh, what are these guys talking about?" loudly enough and a coworker asks who you're talking to, perhaps you should call it a day and read Chapter 8 tomorrow.

A Minor Proclamation

Here's a good place for us to proclaim that you will most likely spend about 10 to 30 percent of your phase logic design effort designing the *Running* logic. The remaining time will be dedicated to designing the *Holding, Restarting, Stopping,* and *Aborting* logic.

When a phase starts, the *Running* logic is executed. If the phase is commanded to *Hold* for any reason (by an operator or by an exception), your batch management package stops executing the *Running* logic and begins running the *Holding* logic. *Holding* logic is responsible for putting the phase into a safe and stable state but is capable of continuing the batch. When the batch is ready to resume, the operator hits a *Restart* button, and the batch management package begins executing the *Restarting* logic. When the *Restarting* logic is complete, control is returned to the *Running* logic and typically resumes where it left off. Figure 11.2 shows *Holding* logic that could work with our example of *Running* logic in Figure 11.1. Note how each control step number is prefixed with an *H*. This signifies *Holding* logic. (In designing the Ben & Jerry's batching system, we prefixed our *Running* logic control steps with an *R*. We didn't put *R*s in Figure 11.1, however, because we wanted to spring the five separate sections of logic on you.)

Figure 11.2 Holding Logic for *CHARGE_INGREDIENT* Phase

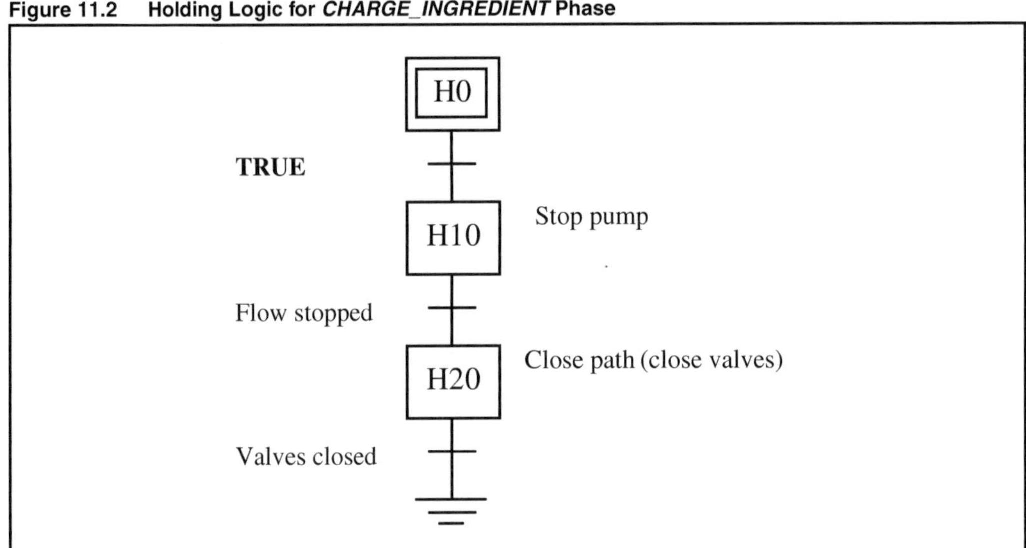

Notice that all the *Holding* logic does is stop the pump and close the valves. It does not download parameters, nor does it upload reports. Its purpose is to place the phase in a safe and stable state. Caution: this assumes, of course, that other phases are not dependent upon this phase for the correct operation or safety of operators, equipment, or product. For example, if your batching process involves transferring two ingredients that must be mixed in some proportion to prevent a reactor from blowing up, holding only one phase *may not be in your best interest.*

Fortunately, OpenBatch and other packages provide an option for propagating state changes. In this example, if a phase goes into the *Holding* state (for any reason), users can choose to have the batch management package automatically hold just that phase, hold all phases within the corresponding operation, hold all phases and all operations within the corresponding unit procedure, or hold the entire recipe.

Even though a disproportionate amount of cream and eggs wouldn't create a runaway exothermic reaction, our operators at Ben & Jerry's asked that we hold the entire batch if a single phase went into *Hold*. They just wanted to know right away what functions were not running and why. So, if one phase has a flow failure it transitions into the *Holding* state and begins running its *Holding* logic. When OpenBatch recognizes this transition, it commands all remaining active phases to transition to their *Holding* states (and thus begin running their respective *Holding* logic).

As shown by Figure 8.1, after the *Holding* logic is complete the phase makes a transition from *Holding* to *Held*. When things are ready and you can resume running the batch, the operator issues a **RESTART** command that begins executing the *Restarting* code. When the *Restarting* logic completes, the phase resumes executing its *Running* code.

Batch management packages remember exactly where in the *Running* logic the batch was placed in *Hold*. Control can return to that exact location after the *Restarting* code is complete. However, you can create placeholders, which are like bookmarks, to tell the batch management package to return to a predefined step instead of the last step executed. For example, let's suppose a phase commands a particular start-up sequence of three pumps that spans 60 seconds. If a **HOLD** command is issued after the first two pumps have started, all pumps stop. If you resumed at the point where you held, you will not restart two of your pumps because the part of the code to start them will not run. Instead, you may choose to resume at the beginning of the start-up sequence so that all the pumps will restart. (Of course, in your *Restarting* code you can add steps to restart the pumps, but, trust us, that will add complexity to your *Restarting* code that you may not want.)

Figure 11.3 shows the *Running* logic for our *CHARGE_INGREDIENT* phase, but this time it includes references to a *HOLD INDEX*. This is OpenBatch's bookmark. The values you assign to the *HOLD INDEX* are arbitrary, but generally the values sequentially increase the further you assign them in the *Running* logic. You can see in Figure 11.3 that *HOLD INDEX* is set to 1 in step R10, to 2 in step R30, and to 3 in step R60.

To see how the *HOLD INDEX* works, Figure 11.4 shows the *Restarting* logic associated with the *CHARGE_INGREDIENT* phase. Remember that the *Restarting* logic is only run when an operator commands a batch in *Hold* to restart.

Pay attention now: notice how the *Restarting* logic steps start with a *T? T* signifies this is *Restarting* logic. We can't have an *R* prefix; that's used in the *Running* logic.

Anyway, as you can see, the first thing the *Restarting* logic checks for is the *HOLD INDEX* value. If it's 2, then the *Restarting* logic forces the *Running* logic execution to resume on step R30. If you look at Figure 11.3, you'll see that R30 and R40 open the valves and start the pump. If we held during step R30, the valves would be open, but the pump would not have started yet. The *Holding* logic closes the valves and stops the pump. (If the pump is not running, no harm is done.) If we attempt to restart at step R40, the pump would start, but the valves would not open!

Application note: at Ben & Jerry's we designed and wrote our PLC phase logic so that the current control step would complete before the phase transitioned to the *Holding* state. This was necessary in order for us to properly control the restarting of the process. If it's critically important that the process be able to go into *Hold* in the middle of a control step, perhaps that control step has too many control actions. Break that control step into two or more separate steps. Back to our example ...

Figure 11.3 Running Logic for the *CHARGE_INGREDIENT* Phase with HOLD INDEX Added

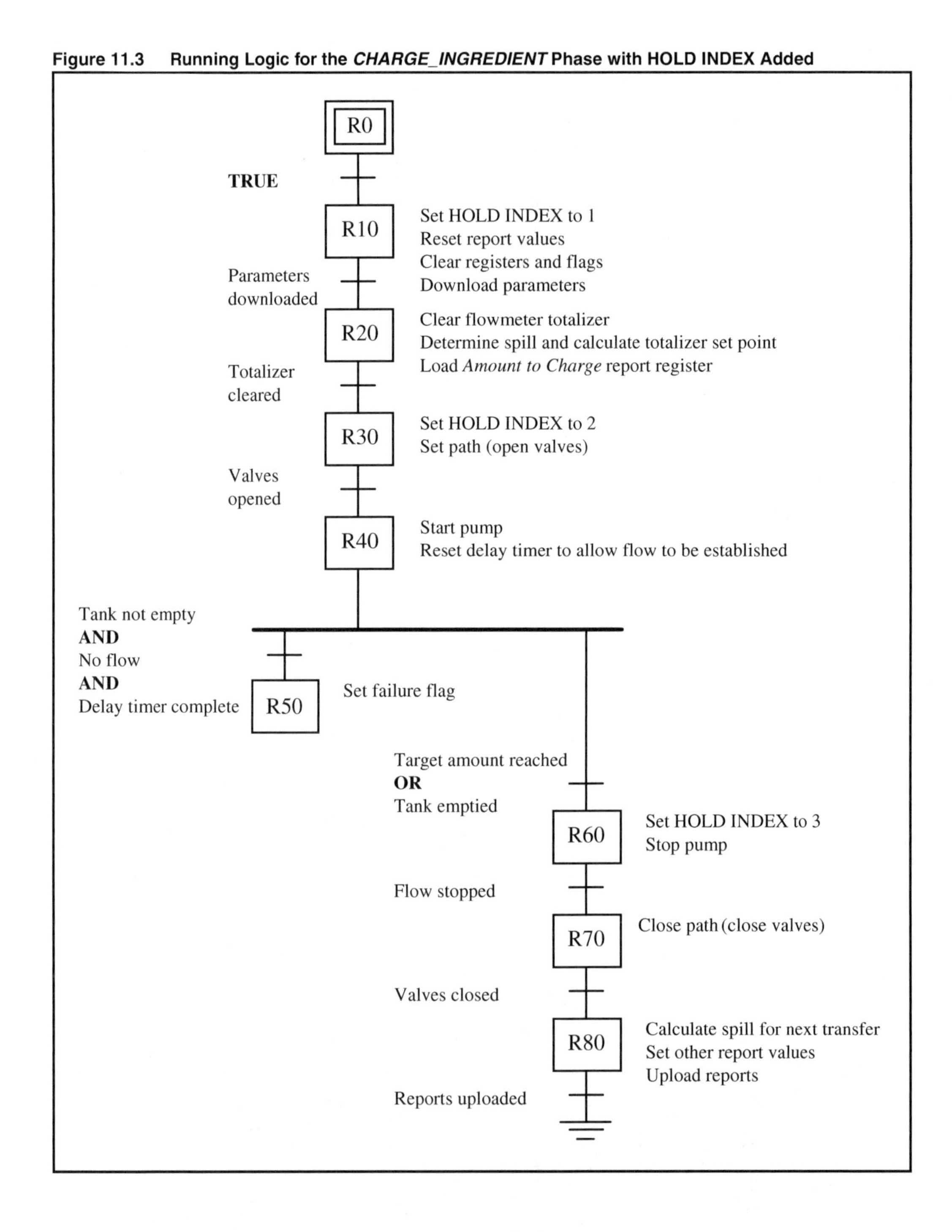

Figure 11.4 Restarting Logic for the *CHARGE_INGREDIENT* Phase

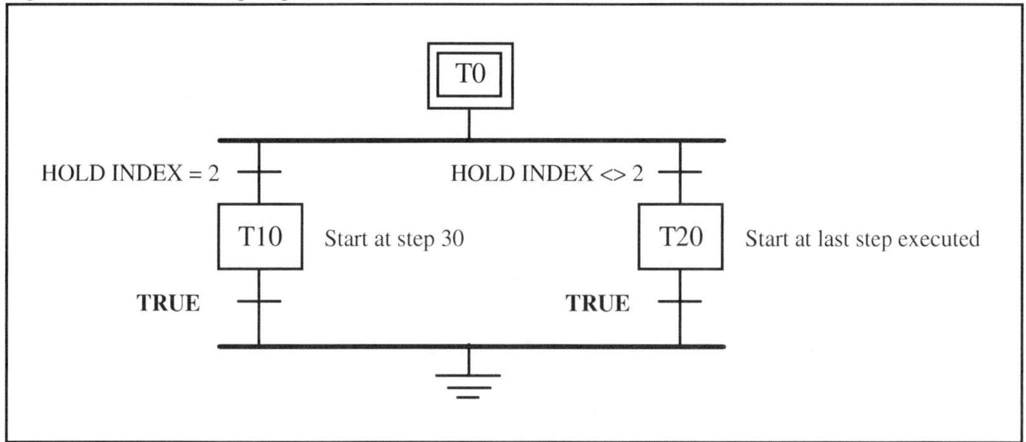

Notice that we also define the *HOLD INDEX* during step R10 and redefine it again in step R60. Essentially, if the phase holds before step R30 or on or after step R60, we want the *Running* logic to resume where it was placed in hold. Deciding where to define *HOLD INDEX* values depends completely on how your process is supposed to run. This can be tricky. At Ben & Jerry's, not defining *HOLD INDEX* values properly was the source of many of our initial phase logic bugs.

Figure 11.5 is the *Stopping* logic for the *CHARGE_INGREDIENT* phase. Notice how it is very similar to the *Holding* logic, except that it adds a step for reporting batch data. Even though you stopped your phase prematurely, you probably still want to record how much ingredient was pumped to the batch tank (or whatever information you are collecting for other phases). According to S88, a *Stopped* phase should not be restarted. So, you probably should collect data while you can.

Keep in mind that *Stopping* logic can be considered the normal termination logic for certain types of phases. For example, the pressure control of a reactor may occur throughout a unit operation. The pressure control phase is stopped only when another event causes it to stop, such as when a particular temperature profile for the reactor is achieved. Executing *Stopping* logic was the normal termination for our *Agitate* phase in making mix. See Figure 5.8.

Okay, *stop*! We know what you're thinking: "Hey, I could create a subroutine and have the *Holding* logic and *Stopping* logic call it. It will save me a lot of programming time and a lot of memory!"

Wrong, wrong, WRONG! Bad idea.

Figure 11.5 Stopping Logic for the *CHARGE_INGREDIENT* Phase

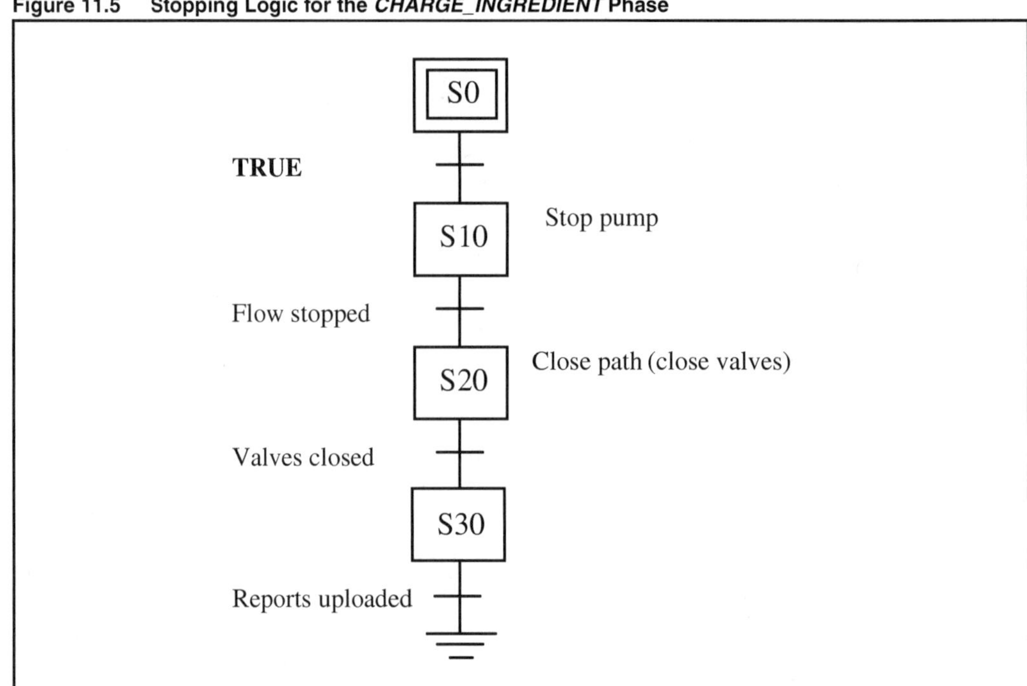

We learned this one the hard way. Yes, you'll be a hero and save a few bytes or words, and perhaps a few minutes of your time, but somewhere, someday, you'll need to modify the way in which a batch holds. You won't want to modify how it stops, and you'll be forced to separate the code anyway. Or worse, you'll modify the common code but forget to validate that it works for both *Holding* and *Stopping*. All programming software packages these days have copy/paste functions that can reduce your programming time. And what about saving some memory? Come on! It was the attempt to save memory that got us into the year 2000 problem! The 1960s-era programmers (who were all retired by the time we Gen-X'ers inherited their work) had good intentions, but it was a big mistake nonetheless.

The *Aborting* logic for the *CHARGE_INGREDIENT* phase is shown in Figure 11.6. A main difference between *Stopping* and *Aborting* can be the speed at which the process halts. *Aborting* may also wish to perform other functions, like dumping the batching material to a drain (or some safe, explosion-proof room). Figure 11.6 halts the *CHARGE_INGREDIENT* phase quicker than its *Stopping* counterpart, but not as cleanly. Look closely at the transitions.

Figure 11.6 Aborting Logic for the *CHARGE_INGREDIENT* Phase

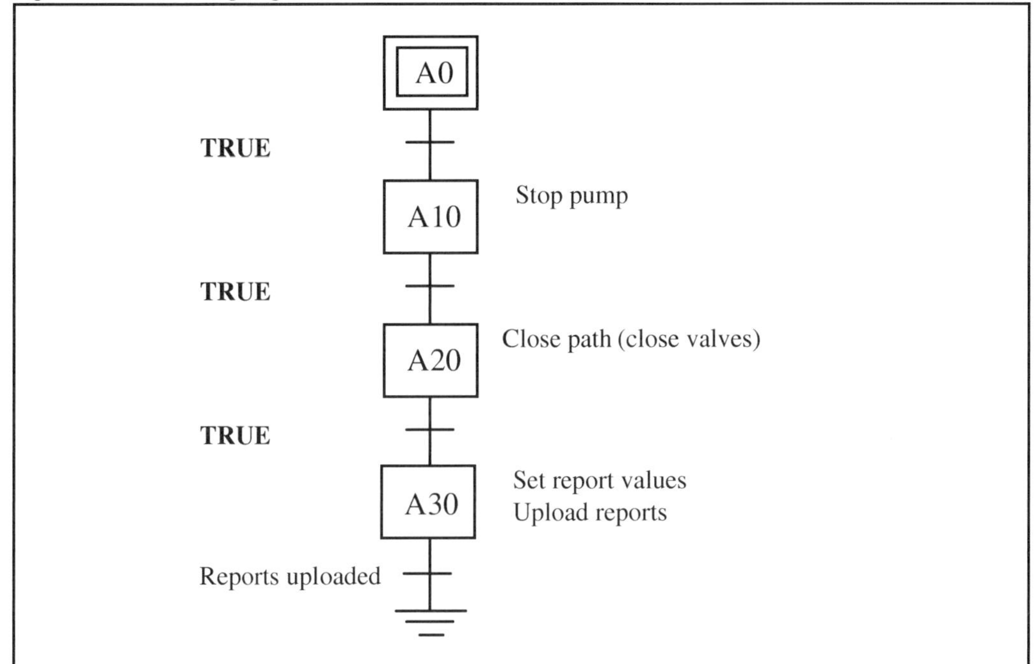

Allocation and Arbitration

Packages such as OpenBatch can handle a lot of equipment allocation and arbitration for you. You assign equipment resource tags (again, watch those tags if you're on a limited budget) to one or more phases. If one phase allocates an exclusive-use resource, any other phase that needs that resource waits until that resource is available before proceeding. Of course, you can explicitly allocate and arbitrate resources within phase logic. It's your choice, depending upon your needs.

Unit-to-Unit Synchronization

You need unit-to-unit synchronization when two units must coordinate functions. Transferring material from one unit to a second unit is a good example of this, but sometimes two units need to transfer their respective materials to a third unit simultaneously. This function can be critical, especially in cases where two volatile chemicals must be mixed in proportion. Packages like OpenBatch serve as a traffic cop with phases, and they help synchronize phases through handshaking with the batch control server.

For example, let's say unit 1 needs to transfer material to unit 2. Unit 1 has finished processing, but unit 2 is not quite empty from the previous batch. Unit 1's *TRANSFER_OUT* phase sends a message to the batch package saying it's ready for the transfer. The batch package will wait until unit 2's *TRANSFER_IN* phase sends a message saying it's ready. Then the batch package will simultaneously (well, as close as it can get) send a "proceed" message to both phases. Voilà! The phases are synchronized. Figure 11.7 shows the *TRANSFER_OUT* phase for unit 1. Figure 11.8 shows the corresponding *TRANSFER_IN* phase for unit 2.

The way these phases are designed, either unit can be ready for the transfer first. In both cases, each phase sends a message to the batch package in step R20, indicating that it is ready for the transfer. When the batch package gets a *Ready* message from both phases, it replies with a *Proceed* message to both phases. This reply triggers the *Transfer Confirmation Received* transition, which allows step R30 to run. (Note that in this example, the *Ready* message is sent by both phases in step R20. The step numbers do not have to be identical in the different phases for a unit-to-unit synchronization to occur.)

Figure 11.7 Example of *TRANSFER_OUT* Phase for Unit 1

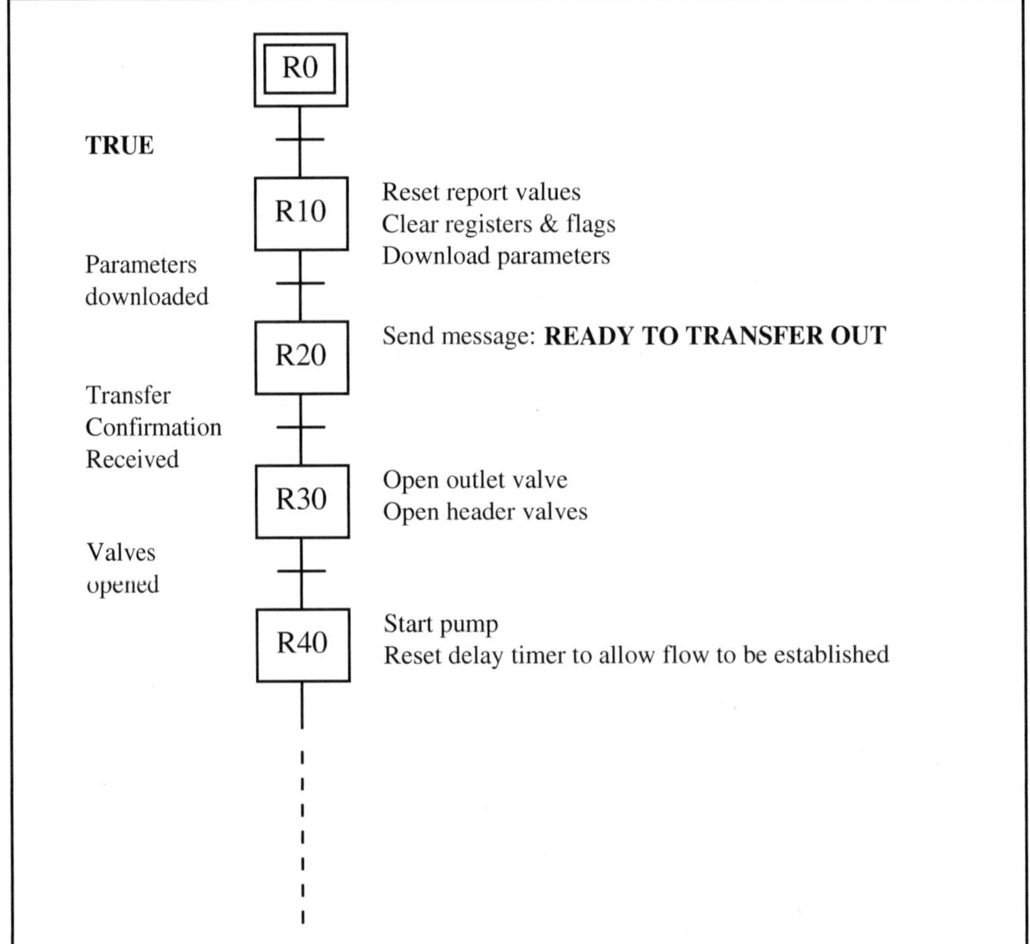

Exception Handling

Phases can deal with exceptions well. OpenBatch has a *Failure* register for each phase, which can be set by phase logic, equipment module logic, or control module logic. The different fail codes are defined when you configure OpenBatch. For example, throughout our mix-making area at Ben & Jerry's, we had yellow "control stop" buttons. These were in addition to hardwired emergency stops that disabled all control power. When these yellow control stop buttons were pressed, they executed shutdown code in the PLC. For the batching part of our control system, logic in each phase constantly monitored the input of each of these buttons. We defined fail code 1 to be *CONTROL_STOP_PRESSED*. If a control stop button was pressed, each phase set its *Failure* register to 1. OpenBatch would sense the transition, command each running phase to *Holding* (and subsequently propagate the entire batch to the *Held* state), and signal the operator that a control stop had been pressed. As another example, in phases where we used flowmeters, we defined fail code 2 to be *FLOW_FAILURE*. If one of our equipment modules detected no flow while a phase was running, it set the phase *Failure* register to 2. OpenBatch would sense this transition, command the phase to *Holding* (and propagate the rest of the batch to *Held*), and signal the operator that a flow failure had occurred.

Figure 11.8 Example of *TRANSFER_IN* Phase for Unit 2

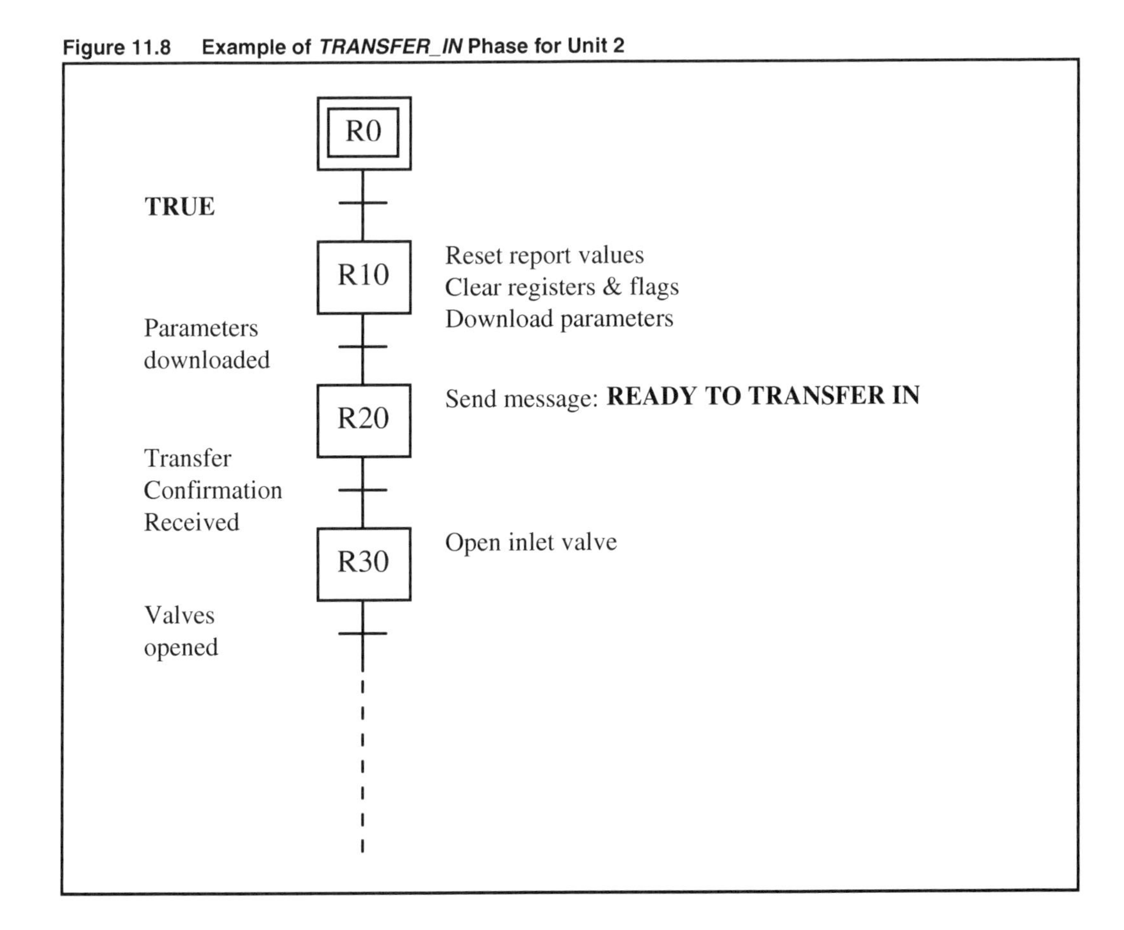

Data Collection

Finally, when specifying the system, keep in mind the type of data you wish to collect. Phase logic can trigger report uploads to batch management packages that can be included in the event journal. As we have mentioned before, you may wish to upload actual quantities used and what source (such as a particular tank) the material came from.

Important Design Notes

Remember that S88 is intended to be used in all systems, from the most manually driven to those requiring the most complicated automation. If designed correctly, the same phase logic used to run a fully automated batch recipe can be used to make a batch in operator-controlled manual steps. So if you work for a global organization, a phase design should reflect the procedure that is needed to make a particular part of a batch. How much you automate depends on your specific site needs.

Sequential function charts can hold much more information than we show in this chapter. Instead of simply saying "Start pump" you can say "Start pump PP1202 (B3/100)," where *PP1202* is the pump ID and *B3/100* is the PLC address that enables the pump. If you are designing a system from scratch, you have the luxury of choosing your own symbols and addresses. Congratulations, but don't screw up by not having standards for symbol names and addresses. An easy thing to do is to use spreadsheets to allocate addresses for such things as registers, timers, and control bits. If you want to be more elaborate in your documentation, include what writes to and reads from each address and why.

More than likely, you're going to be managing or working on a project with several people. Be careful who "owns" the master copies of documents. Use librarian software or assign a person to be the master keeper. Do whatever works for you, but understand who owns the documents. At Ben & Jerry's, we didn't have any fancy librarian software. For our SFCs, we had a box with the word *MASTER* on the cover page for each phase. On the master copy, we colored this box in with a yellow highlighter. An engineer was not allowed to make changes to a phase unless he had the copy with the yellowed master box.

To speed things along with addresses, we would preallocate blocks of addresses to each engineer. For example, Larry was allowed to use timers T4:100 through T4:109 for a phase he was working on. You have to guess well doing this, however, or you may end up with a lot of unused addresses wedged in between used addresses.

Congratulations, you made it through the longest chapter. Go take a break.

12

Writing Phase Logic

We would not consider this book complete without an actual discussion about writing code. However, you'll notice a distinct favoritism on our part toward PLCs (versus DCSs) in this chapter. There are four good reasons for this:

- First, modern-day distributed control systems often use sequential function charts (SFCs) as a programming language. Therefore, translating phase logic design (using SFCs) into executable code is more straightforward in a DCS environment than it is in a ladder-logic-based PLC environment. (In other words, we don't feel we need to spend as much time on DCS phase logic.)
- Along those same lines, since distributed control systems were designed with the process industries in mind and since a majority of S88 committee members used (and use) distributed control systems extensively, it's no surprise that DCSs are better prepared to implement solutions using S88 models and terminology. Therefore, DCS users tend to already know about things like control modules, equipment modules, and units. They may just not be accustomed to working with them within the context of S88.
- Third, configuring a system (versus writing phase logic) represents a significant amount of effort when you are implementing S88 using a DCS. Since configuration is unique to the particular DCS that you're using, we felt it was not in your best interests to add one hundred pages of configuration material to this book.
- Next, this is a practical book about applying S88. Since our solution at Ben & Jerry's was PLC-based, we're going to limit our DCS discussion to an important overview about working DCS phase logic.

Using Distributed Control Systems

As we said earlier, distributed control systems have been used for batch control since someone decided control systems should be distributed. Seriously, remember that PLCs were created to make cars. DCSs were created to make chemicals and refine gasoline. Therefore, even early distributed control systems had the architecture and infrastructure to handle batch and continuous manufacturing. PLCs were more adept at controlling discrete manufacturing.

Although over the past thirty years other languages have been developed for PLCs, ladder logic is still the most popular method of programming these systems. Distributed control systems have used various scripting or graphical languages, as outlined in the IEC-61131-3 programming language standard. Two popular DCS programming languages include sequential function charts (SFCs) and structured text. Clearly, if SFCs are used as a programming language, translating your SFC-based design into executable code shouldn't take too much effort.

In recent years, hybrid DCS systems have evolved that combine the modularity and relative low cost of PLCs with the integration robustness of a traditional DCS. Fisher-Rosemount's DeltaV, Foxboro's Fox-IA, and Honeywell's Plantscape are examples of hybrid systems. Adding credibility to the hybrid architecture, Siemens has its PCS 7 system and Rockwell recently introduced its ProcessLogix solution. (Honeywell's Plantscape and Rockwell's ProcessLogix are essentially the same system in that the two companies codeveloped it using Honeywell's software and Rockwell's ControlLogix hardware.)

Each of these hybrid solutions allows you to define phase logic using sequential function charts. Figure 12.1 is *Running* logic from a phase called *CHARGE_SWEETENER*. (This is the same as Figure 11.4, only we called it *CHARGE_INGREDIENT* back there.) While the SFC editors in each hybrid product may make the SFC appear differently, the essential step and transition logic is the same.

Figure 12.1 ***CHARGE_SWEETENER*** **Phase Running Logic**

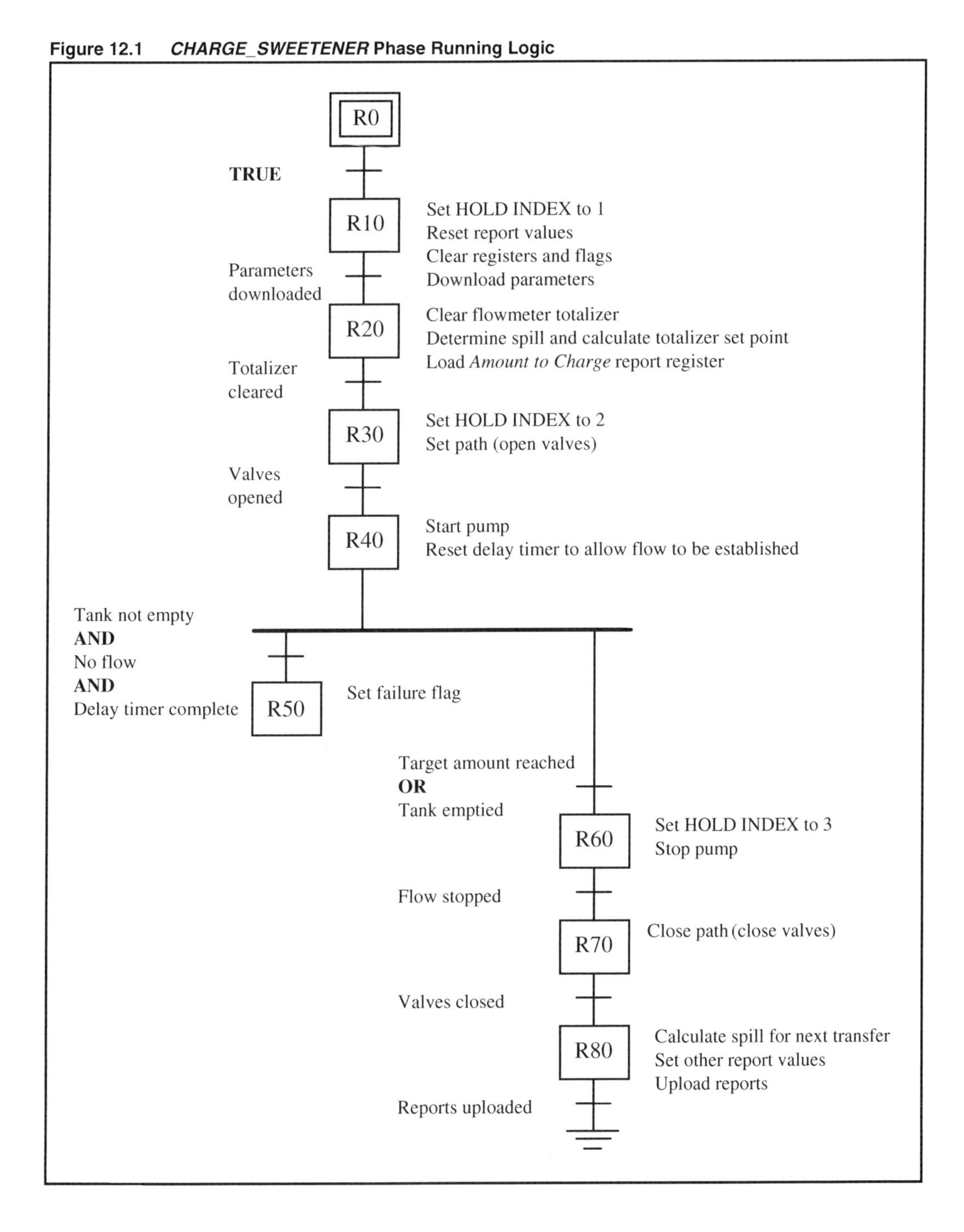

Writing PLC Phase Logic

Figure 12.2 is the Allen-Bradley PLC-5 ladder logic for step R10 of the *CHARGE_SWEETENER* phase (Figure 12.1). The *EQU* instruction performs an equate. If A=B, the instruction is true and the rung executes. The *MOV* instruction moves the contents of the source into the destination. The first branch of Figure 12.2 moves the constant 10 into the integer register N32:3. (In Allen-Bradleyspeak, *N* means integer, as *I* means input; so file 32 is an integer file, and we are transferring the constant 10 into location 3 of that file.) N32:3 holds the step number for the *CHARGE_SWEETENER* phase. Putting 10 into this register tells other parts of the PLC code and the batch management system that the phase is in step 10. The third branch of Figure 12.2 has a fill file (*FLL*) instruction that fills ten registers with zero, starting at N105:31.

You'll need to notice a few important things about Figure 12.2. First, all of step 10 is on one "rung" of the ladder. We commented on this in Chapter 11. This is how we chose to implement our code; S88 does not require this. (S88 does not even call out that you need to use PLCs or DCSs.) Typically, PLC programmers like looking at an entire rung at a time on their computer screen. This may be difficult with some phase logic rungs because there are so many branches. How you implement your ladder logic is unimportant. Consistency and maintainability are vital. (For us, readability at 2:00 a.m. was also vital for those moments when the Ben & Jerry's maintenance staff called us at home to help them track down a problem.)

Second, if you refer back to Figure 12.1, the transition condition required to move from step R0 to step R10 is simply *TRUE*. Therefore, once we're in step R0, we immediately transition to step R10.

Third, do you see how this rung has an inherent "one-shot" logic? In order to execute the rung, *SWEET_SI* (N32:3) needs to be equal to 0. However, the first branch in the rung sets *SWEET_SI* to 10, so this rung won't run during the next PLC program scan. *SWEET_SI* is the sweetener phase step index. This PLC register holds the value of the phase step (such as R10 and R20). Because of this one-shot logic, when you are manipulating discrete outputs you'll probably want to use *latch* and *unlatch* output instructions, `-(L)-` and `-(U)-`, instead of a momentary output instruction, `-( )-`. (If you use a momentary output, you'll only open valves or start pumps for one scan.)

A word of caution: keep in mind that latching functions have safety implications. If power is lost and comes back on, the system can come up with valves open, pumps running, and so on. Make sure you account for this. In our PLC code, whenever the PLC first-scan bit was set (indicating that the PLC was just powered up or just placed in *Run* mode) our code disabled all latched enable bits.

Figure 12.2 ***CHARGE_SWEETENER*** **Phase Running Logic for Step R10**

```
   |Sweetener                                             Sweetener             |
   |Phase                                                 Phase                 |
   |Step Index                                            Step Index            |
   |Register                                              Register              |
   |SWEET_SI                                              SWEET_SI              |
   |+--EQU-----------+                                    +--MOV-----------+    |
 32++Equal (A=B)     +----------------------------------++Move            +-+-+
   ||A:         N32:3|                                  ||Source:       10| | |
   ||B:             0|                                  ||                | | |
   |+----------------+                                  ||Dest:      N32:3| | |
   |                                                    ||               0| | |
   |                                                    |+----------------+ | |
   |                                                    |Sweetener          | |
   |                                                    |Phase              | |
   |                                                    |Hold Index         | |
   |                                                    |Register           | |
   |                                                    |SWEET_HI           | |
   |                                                    |+--MOV-----------+ | |
   |                                                    ++Move            +-+ |
   |                                                    ||Source:        1| | |
   |                                                    ||Dest:      N31:3| | |
   |                                                    ||               6| | |
   |                                                    |+----------------+ | |
   |                                                    |Sweetener          | |
   |                                                    |Phase              | |
   |                                                    |Amount to          | |
   |                                                    |Charge             | |
   |                                                    |SWEETR01           | |
   |                                                    |+--FLL-----------+ | |
   |                                                    ++Fill File       +-+ |
   |                                                    ||Source:        0| | |
   |                                                    ||Dest:   #N105:31| | |
   |                                                    ||Length:       10| | |
   |                                                    |+----------------+ | |
   |                                                    |Sweetener          | |
   |                                                    |Spill Storage      | |
   |                                                    |Address            | |
   |                                                    |Indirection        | |
   |                                                    |Index              | |
   |                                                    |+--MOV-----------+ | |
   |                                                    ++Move            +-+ |
   |                                                    ||Source:        0| | |
   |                                                    ||Dest:    N100:59| | |
   |                                                    ||              57| | |
   |                                                    |+----------------+ | |
   |                                                    |Sweetener          | |
   |                                                    |Phase              | |
   |                                                    |Request            | |
   |                                                    |Register           | |
   |                                                    |SWEET_RQ           | |
   |                                                    |+--MOV-----------+ | |
   |                                                    ++Move            +-+ |
   |                                                     |Source:     1000|   |
   |                                                     |Dest:     N26:18|   |
   |                                                     |               0|   |
   |                                                     +----------------+   |
```

Finally, notice that the step is only a numeric value. There is no *R* preceding it. (The PLC *MOV* instruction only works with numbers.) So how does the batch management package and the PLC differentiate between step 10 in *Running* logic and, let's say, step 10 in *Holding* logic? That's easy, since there's a running active bit for each phase that OpenBatch sets whenever it wants the phase's *Running* logic to execute. (In this example, we'll use the symbol *SWEET_R* and the PLC address N28:3/4. There are also bits for the other phase state logic: *SWEET_A*, *SWEET_S*, *SWEET_H*, and *SWEET_T*.) There are two ways of using this bit to control which state logic you wish to run. You can add an examine-if-closed (XIC) bit to each rung for *SWEET_R*. Alternatively, for Allen-Bradley PLC users, you can use the **MCR** command, as shown in Figure 12.3.

Figure 12.3 Running Logic MCR Rung

```
  |Sweetener                                                              |
  |Phase                                                                  |
  |Running                                                                |
  |Active                                                                 |
  |SWEET_R                                                                |
  |    N28:3                                                              |
31+----] [------------------------------------------------------------[MCR]+
  |      4                                                                |


                         (Running logic code in here)


  |                                                                       |
54+-------------------------------------------------------------------[MCR]+
  |                                                                       |
```

When the condition or conditions to enable the **MCR** are true, all code between the two **MCR** rungs is allowed to run. If the **MCR** is not enabled, the code between the two **MCR** rungs is not allowed to run.

Figure 12.4 shows how you can implement step R50. Figure 12.5 shows how you might implement the conditions to transition into step R60, and Figure 12.6 shows how you might implement the actions of step R60.

Figure 12.4 ***CHARGE_SWEETENER*** **Phase Running Logic Step R50**

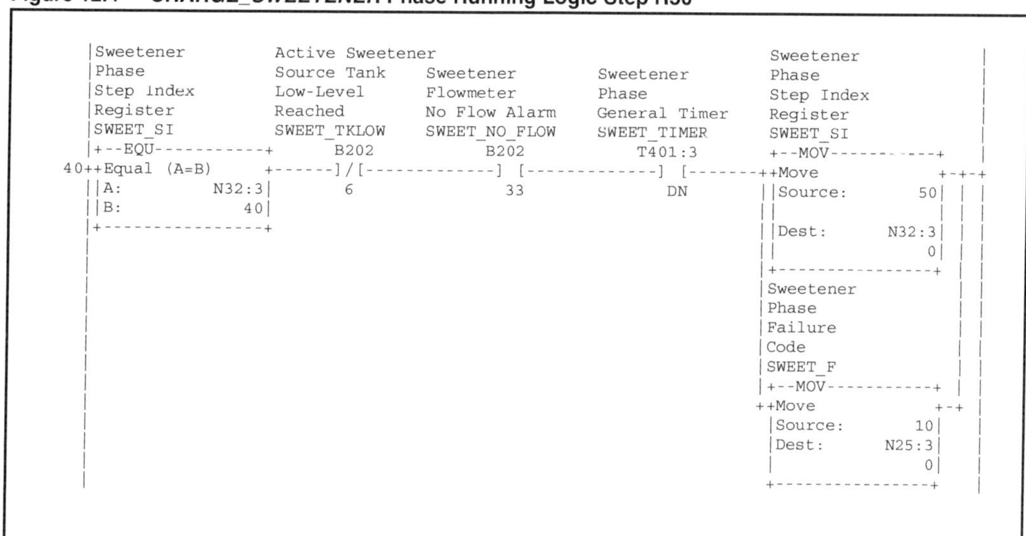

Figure 12.5 Conditions to Transition into Step R60

```
  | Sweetener
  | Phase
  | Step Index           Sweetener
  | Register             Flowmeter Total
  | SWEET_SI             SWEET_PV
  | +--EQU-----------+   +--GEQ--------------------+
46|++Equal (A=B)     +--++Grtr Than or Equal (A>=B)+----------------------------------------+--
  | |A:        N32:3|   ||A:              N100:55|                                          |
  | |B:           40|   ||B:              N100:54|                                          |
  | +---------------+   |+-----------------------+                                          |
  |                     |Active Sweetener                                                   |
  |                     |Source Tank    Sweetener                            Sweetener      |
  |                     |Low-Level      Flowmeter       Sweetener            Phase          |
  |                     |Reached        No Flow Alarm   Flowmeter Total      General Timer  |
  |                     |SWEET_TKLOW    SWEET_NO_FLOW   SWEET_PV             SWEET_TIMER    |
  |                     |    B202           B202      +--LES-----------+       T401:3       |
  |                     +------] [-------------] [-------+Less Than (A<B) +-------] [------+
  |                          6              33        |A:      N100:55|        DN
  |                                                   |B:      N100:54|
  |                                                   +---------------+
```

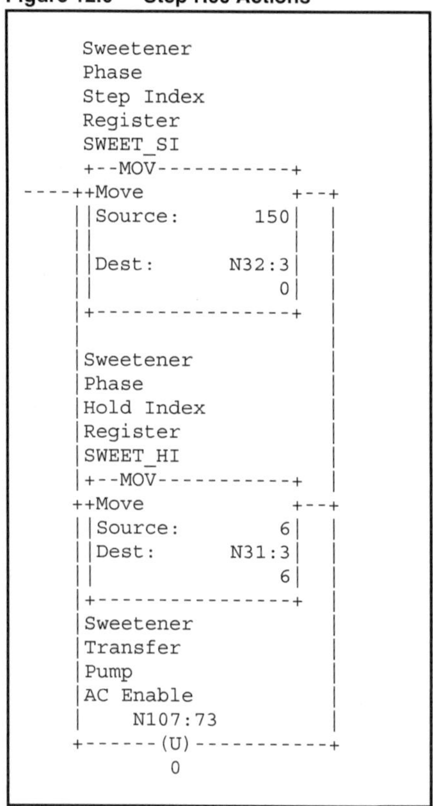

Figure 12.6 Step R60 Actions

The inherent one-shot property of this type of phase logic is very convenient for procedural control. That is, in your procedural code you don't want every rung running on every scan. However, there are certain conditions you do want checked during every scan, such as flow rates, totalizer values, and tank levels. Also, in the case of charging sweetener, at Ben & Jerry's we have more than one sweetener tank, each of which has its own flowmeter. Having the phase logic check which tank and flowmeter to use during every control step adds an unmanageable amount of code. Instead, we created a separate PLC file to handle control items like these. Let's talk about this now.

In the ladder logic for steps R50 and R60, you might have noticed bits and words labeled *SWEET_TKLOW*, *SWEET_NO_FLOW*, and *SWEET_PV*. These are the sweetener tank empty condition, sweetener no-flow condition, and sweetener flowmeter totalizer values, respectively. Figure 12.7 shows the equipment code for handling the tank empty condition. This code is kept in a separate PLC file and is not part of the phase logic.

Figure 12.7 Equipment Code to Handle Tank Empty Condition

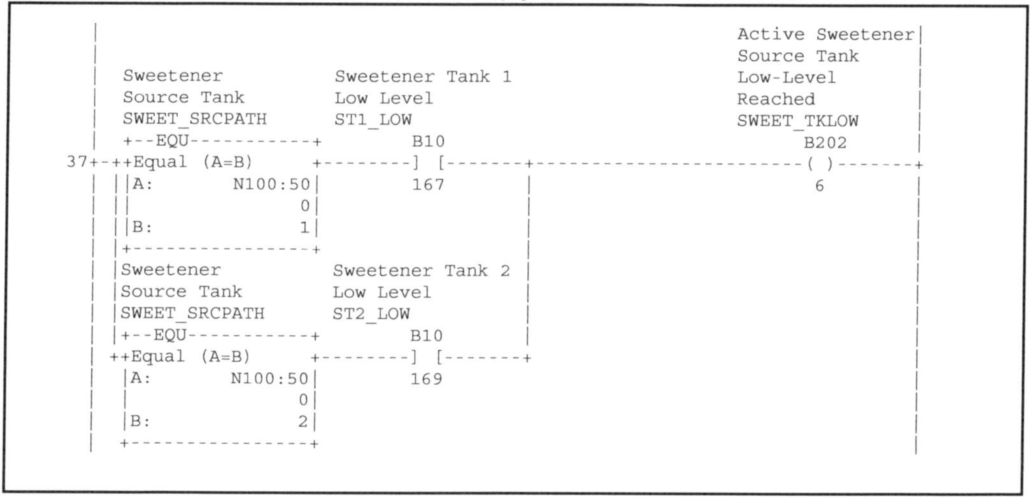

```
     |                                                           Active Sweetener|
     |                                                           Source Tank     |
     | Sweetener               Sweetener Tank 1                  Low-Level       |
     | Source Tank             Low Level                         Reached         |
     | SWEET_SRCPATH           ST1_LOW                           SWEET_TKLOW     |
     | +--EQU-----------+             B10                              B202      |
  37+-++Equal (A=B)     +--------] [-------+----------------------( )-------+
     | ||A:        N100:50|           167         |                       6          |
     | ||               0|                        |                                  |
     | ||B:              1|                       |                                  |
     | |+----------------+                        |                                  |
     | |Sweetener              Sweetener Tank 2   |                                  |
     | |Source Tank            Low Level          |                                  |
     | |SWEET_SRCPATH          ST2_LOW            |                                  |
     | |+--EQU-----------+            B10         |                                  |
     | ++Equal (A=B)     +--------] [-------+                                        |
     |  |A:        N100:50|          169                                             |
     |  |               0|                                                           |
     |  |B:              2|                                                          |
     |  +----------------+                                                           |
```

Notice how we use *SWEET_SRCPATH* (sweetener source path) combined with a bit that indicates a tank is low to set or reset *SWEET_TKLOW*. *ST1_LOW* and *ST2_LOW* are the result of the tank low-level discrete input being run through a debounce timer. *SWEET_SRCPATH* is set in R30 based on a parameter that was downloaded when the phase began. *SWEET_TKLOW* is used in R50 and R60.

Figure 12.8 shows the code that handles the flowmeters that are associated with both sweetener tanks.

Notice how the flowmeter equipment code uses *SWEET_SRCPATH* to choose which flowmeter data to move into *SWEET_PV* (N100:55) and *SWEET_GPM* (N100:56). (An Allen-Bradley flowmeter I/O module loads scaled values directly into the N24:16, N24:18, N24:25, and N24:27 registers.) Also notice how *SWEET_GPM* is used to trigger the *SWEET_NO_FLOW* condition.

Figure 12.8 Equipment Code to Handle Sweetener Flowmeters

```
     |Sweetener                                             Sweetener          |
     |Source Tank                                           Flowmeter Total    |
     |SWEET_SRCPATH                                         SWEET_PV           |
     |+--EQU-----------+                                    +--MOV-----------+ |
   39++Equal (A=B)     +-----------------------------------++Move           +-+-+
     ||A:        N100:50|                                  ||Source:   N24:18| | |
     ||                0|                                  ||             553| | |
     ||B:              1|                                  ||Dest:    N100:55| | |
     ||                 |                                  ||             552| | |
     |+-----------------+                                  |+----------------+ | |
     |                                                     |Sweetener          | |
     |                                                     |Flowmeter Flow     | |
     |                                                     |Rate (GPM)         | |
     |                                                     |SWEET_GPM          | |
     |                                                     |+--MOV-----------+ | |
     |                                                     ++Move           +-+ |
     |                                                      |Source:   N24:16|   |
     |                                                      |               0|   |
     |                                                      |Dest:    N100:56|   |
     |                                                      |               0|   |
     |                                                      +----------------+   |
     |Sweetener                                             Sweetener          |
     |Source Tank                                           Flowmeter Total    |
     |SWEET_SRCPATH                                         SWEET_PV           |
     |+--EQU-----------+                                    +--MOV-----------+ |
   40++Equal (A=B)     +-----------------------------------++Move           +-+-+
     ||A:        N100:50|                                  ||Source:   N24:27| | |
     ||                0|                                  ||               0| | |
     ||B:              2|                                  ||Dest:    N100:55| | |
     ||                 |                                  ||             552| | |
     |+-----------------+                                  |+----------------+ | |
     |                                                     |Sweetener          | |
     |                                                     |Flowmeter Flow     | |
     |                                                     |Rate (GPM)         | |
     |                                                     |SWEET_GPM          | |
     |                                                     |+--MOV-----------+ | |
     |                                                     ++Move           +-+ |
     |                                                      |Source:   N24:25|   |
     |                                                      |               0|   |
     |                                                      |Dest:    N100:56|   |
     |                                                      |               0|   |
     |                                                      +----------------+   |
     |Sweetener                                 Sweetener                         |
     |Flowmeter Flow                            No Flow Test                      |
     |Rate (GPM)                                Delay Timer                       |
     |SWEET_GPM                                                                   |
     |+--LES-----------+                        +--TON-----------+                |
   42++Less Than (A<B) +-----------------------++Timer On Delay +-(EN)---------+-+
     ||A:        N100:56|                      ||Timer:   T401:28|             | |
     ||                0|                      ||Base (SEC):  1.0+-(DN)        | |
     ||B:             10|                      ||Preset:        2|             | |
     ||                 |                      ||Accum:         2|             | |
     |+-----------------+                      |+----------------+             | |
     |                                         |Sweetener      Sweetener       | |
     |                                         |No Flow Test   Flowmeter       | |
     |                                         |Delay Timer    No Flow Alarm   | |
     |                                         |               SWEET_NO_FLOW   | |
     |                                         |  T401:28          B202        | |
     |                                         +-----] [-------------( )-------+ |
     |                                               DN              33          |
```

Writing Control Modules (Device Drivers)

In Chapter 4, we talked about control modules. Remember that a control module is typically a collection of sensors, actuators, other control modules, and associated process equipment that from the point of view of control is operated as a single entity. Each control module provides a direct "connection" to the process through actuators and sensors.

In the simplest form, control modules can just be device drivers (Attention NT bit heads: don't confuse *control system device drivers* as we're discussing them with *NT device drivers*.) Control modules should provide a robust method of device control, including these functions:

- Automatic and manual modes
- Simulation mode
- Permissives
- Alarms

For example, let's say you want to control the rate of flow of some liquid. You use a pump driven by a variable-speed motor and a flowmeter. A PID-type controller reads the flow rate and sets the pump speed appropriately. The batch control system wants only to control the flow rate and considers the combination pump, flowmeter, and PID loop as a control module. Keep in mind that even though the control module exists in the physical model, not all elements must be physical. In our example, the PID controller can be a PLC instruction and not a physical stand-alone device physically linking the flowmeter to the pump.

By the nature of their operation, distributed control systems handle a lot of the work of control modules for you. You can set up a pump, flowmeter, set point, modes, permissives, and alarms all within a predefined block. Programmable logic controllers, however, were designed to make cars. That's right; don't let anyone tell you otherwise: PLCs were designed for discrete manufacturing. In recent years, vendors have marketed lots of PLC modules for process control and instrumentation. Companies like Ben & Jerry's chose PLCs over a DCS for one very good reason: moola. For us, PLCs had a lower initial and life-cycle cost than a DCS (at least in 1993).

And so, to incorporate nifty little features like modes, permissives, and alarms you'll need to write one or more device drivers. Figure 12.9 (which is identical to Figure 4.8) shows the design of a typical device driver control module. In our system, we created a contiguous block of registers to handle enabling, modes, permissives, and alarms. This way, we could use a loop and indirect addressing to handle every output. We're not going to include the code here, but it's a straightforward application of such operators as *AND* and *OR*.

Figure 12.9 A Sample Device Driver Control Module

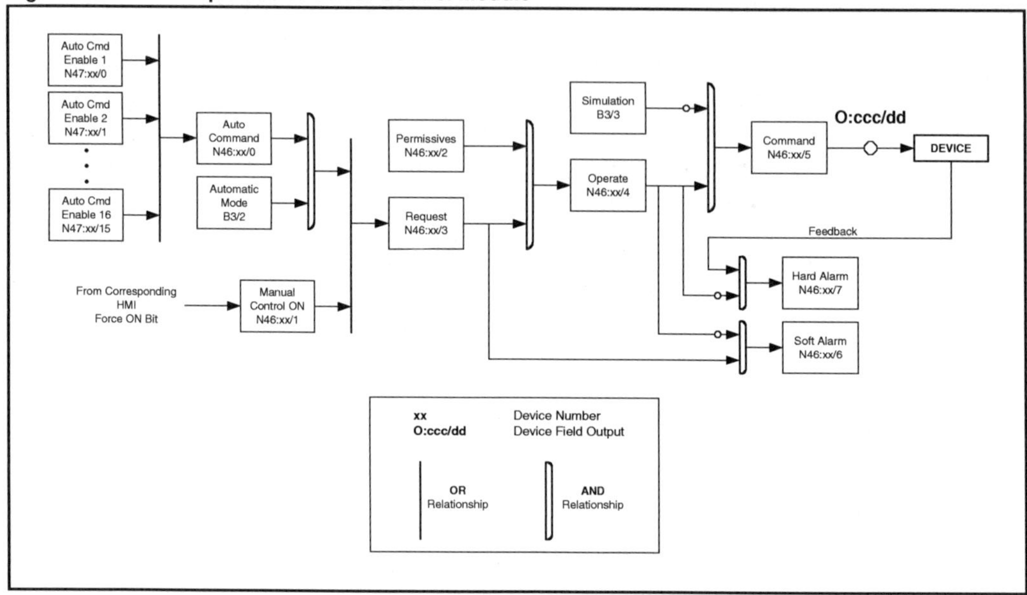

Writing this in ladder logic can be a lengthy and not-so-easy task. Hybrid systems, on the other hand, generally use function blocks to define control modules for each device. Figure 12.10 shows such a control module for a sweetener valve using the Siemens PCS 7 system.

A Design/Code Process

When we set out to design our phase logic at Ben & Jerry's, we took the high road: we started with our most complex phase. Its functionality was clearly a superset of many other phases. This made us think hard. If you don't want to think hard right away, start with your easiest phase. Then, when you consider yourself a genius, go on to your most complex phase and be humbled.

When you finish designing your first phase, consider writing the code for it. You'll probably learn a few things while you're coding. Update your phase design and then move on to your next phase. Maybe code that when you're done, then design the rest of your phases. And, hey, this method works regardless of whether you're using PLCs or a DCS!

Figure 12.10 A Sample Control Module Configuration Using a Siemens PCS 7 System

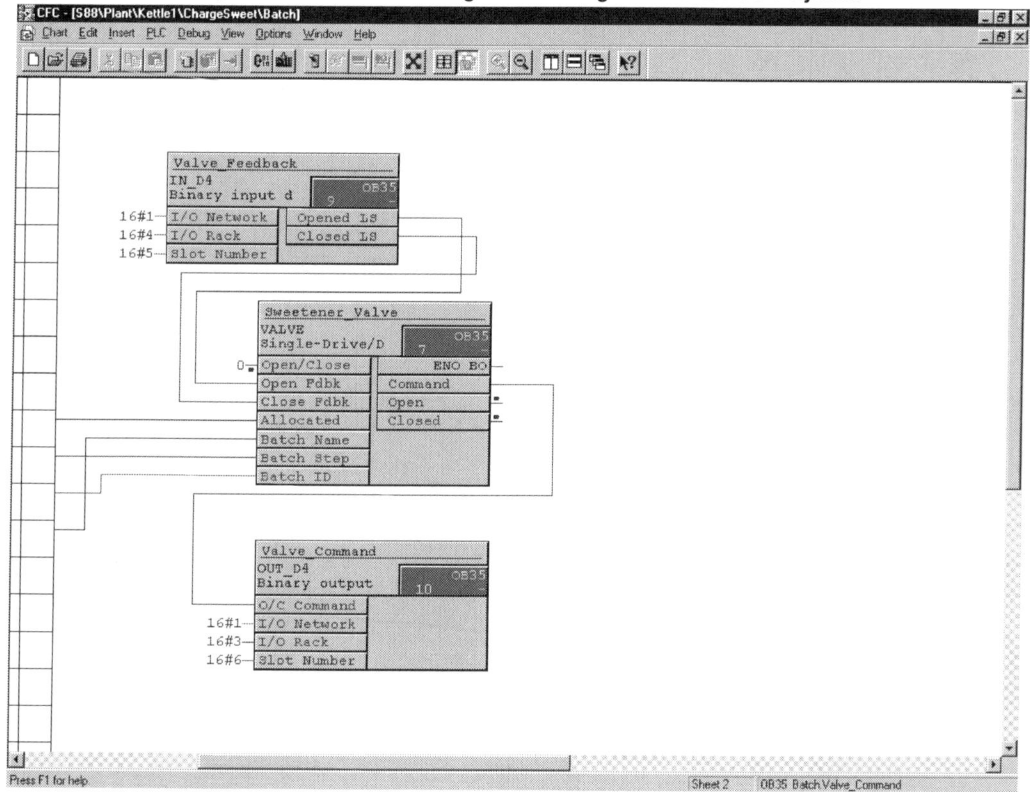

TIPS

When you are creating data space in your PLC, create some extra. When working with Allen-Bradley PLC-5s, you cannot create more program files or data space if you are in *Run* or *Test* mode. The PLC we used for our batching system also handled things like receiving and CIP. Just because we weren't batching didn't mean we could throw the PLC into *Program* mode (essentially shutting it down).

Going along with that thought, create a few spare phases. No matter how thoroughly you designed the system, in the middle of testing you'll realize that you need an extra phase for one thing or another. Maybe you require a simple delay phase.

Depending on how you assign phase parameters and reports to PLC registers, you may want to create a few spare parameters or reports per phase. If phase 1 requires three parameters and phase 2 requires four parameters, you could assign phase 1 parameters to N401:1-3 and phase 2 parameters to N401:4-7. However, if you later discover that you needed a fourth parameter on phase 1, you either have to live with noncontiguous parameter addresses or reassign all your parameters. Careful though: parameters and reports are tags.

The PLI

So far, we've been talking about the actual equipment phase code. Normally, there is different code that interfaces with a batch management package to interpret commands, send requests, deal with state transitions, and handle situations when communication is lost between a PLC or DCS and the batch management package. This is called the *phase logic interface* (PLI). In the olden days, PLIs had to be written for every S88 project. Modern technology now allows batch management vendors (or their OEMs) to include suggested PLIs with the batch management software. You may feel we are stating the obvious here, but each PLC will have its own PLI.

Dealing with lost communication between a PLC or DCS and the batch management package is a critical function of the PLI. A configurable watchdog timer or heartbeat is often used to let the systems know that they are successfully communicating with each other. When the communication is lost, perhaps because of a broken network connection or a crashed personal computer, the PLI can put all active phases in *Hold*. If the entire control system loses power, the PLI can also be configured to place all previously active phases in *Hold* on start-up. (Depending on your safety or quality needs, you may wish to reset all previously active phases to the *Idle* state.) Most batch management packages on the market are robust enough to remember what your recipe was doing before a power loss, so there is a decent chance you will be able to recover and save your batch. (If that is what you want to do. Again, safety or quality factors may dictate another course of action.)

Like control modules, PLIs in hybrid systems are defined with function blocks. For PLCs, the PLI is often created with ladder logic. The PLI code makes use of specific PLC files and registers that are normally reserved for its exclusive use. Since we were retrofitting our S88 solution into an existing PLC, we were already using about 80 percent of those files and registers. So, we could do the following:

1. Reassign the PLI reserved addresses.
2. Reprogram the existing receiving and clean-in-place (CIP) code.
3. Trash the project.
4. Go have a beer and think about it later.

We did option 4 about three times, and then we decided to reassign the PLI reserved addresses. Fortunately, Allen-Bradley PLC programming software has a nice search-and-replace feature that came in handy.

Now that we've discussed writing code, it's time to talk about how to prove that it's going to work. Ah, validation.

13

Starting Your System Right ... the First Time

The section on starting your system is in Chapter *13*? A coincidence? We don't think so. Superstition aside, we do think it's very appropriate to discuss system start-ups. Call it what you will: validation, commissioning, system acceptance; it all comes down to ensuring that the system does what you want it to do.

Validation

Most companies are just as concerned with the process that creates their products as they are with the products themselves. They will build or buy the best process equipment and systems to satisfy their needs. The U.S. Food and Drug Administration (FDA) requires pharmaceutical companies to perform formal system validation on all automated processes. Petroleum companies perform formal validation because they generally don't want their plants to blow up when starting a new process.

At the ice cream plant, we underwent semiformal validation during the start-up of our batch control system because it simply made business sense. It was semiformal validation because we wore business suits instead of tuxes. Seriously, the food industry does not have the same stringent requirements for validation as our consumer-products cousins making pharmaceuticals. So, we had the luxury of picking and choosing specific validation exercises that best met our needs. Don't scoff at us, you pharmaceutical engineers. It wasn't too long ago that Ben & Jerry's created requirements and design notes on the back of beer napkins. (We're not kidding.) Anyway, our validation included three steps:

1. Simulation
2. Water testing
3. Product testing

Our water testing was loosely analogous to a pharmaceutical operational qualification (OQ). Likewise, our product testing was analogous to a performance qualification (PQ).

We didn't have any fancy simulation software. We loaded the PLC code into a bench PLC and ran each of our equipment phases in manual mode with OpenBatch. We observed the progress of each phase with Allen-Bradley programming software. Simulating all our phases took about ten days, and we found 90 percent of our design flaws or program bugs during this time. To help with simulation, we created a spreadsheet matrix for each phase. Across the top of each matrix was the list of functions we wanted to test for that phase. Down the left side was a list of the batch IDs used to test the phase. (Even in manual mode, OpenBatch requires that a batch ID be entered. So, every time we tested a phase, we created a batch ID identifying a test.)

Because we had to test a couple of dozen different functions for some of our phases, we could not test all of them in one phase execution. Often, we would test different combinations of functions, in different orders. You may face the same challenge. For example, in one run, you may wish to test a flowmeter failure and then a phase *Restart*. In the next run, you may wish to test a flowmeter failure and then a phase *Stop*. You may also wish to test a flowmeter failure with a phase *Abort*. If you have the option to use a primary and secondary tank, you may wish to test the flowmeter failure combinations for both tanks. Only you can decide which tests are relevant. Some of our phases required thirty-one or more tests.

Simulation testing can be very involved and take a lot of time. Not only did we want to ensure that the right valves opened and the right pumps started at the proper time, but we also needed to verify that other valves did not open and other pumps did not start. If we found an error, we would retest the condition.

We tested all functions with simulation. Our simulation testing was so successful that water testing only took six hours on a Saturday. That made Althea, the lead mix maker, very, very happy. It made Eugene, the plant environmental coordinator, happy too. Ben & Jerry's was always focused on the impact that manufacturing had on the environment. The less water we used and the less waste we created, the better. We tested phases, as well as running recipe procedures, during water testing.

Product testing is critical because not all of your products will use all of your recipe procedures the same way. Product testing was especially important to us because we couldn't water test everything. For example, the sweetener transfer lines remained charged all the time. We could not water test those. Since we wanted to minimize start-up time, cost, and waste, product testing was done with planned production batches. So we had to be very alert during actual production runs.

Our first production batch was a success. We won't deny that we had little glitches here and there as we performed more complex functions, but nothing catastrophic.

As we said in Chapter 1, the very nature of S88 modularity helps with validation. If designed properly, equipment phases are functionally independent. And so, once phase A is validated, modifying other phases won't upset phase A's validated state. Also, validating a recipe procedure is easier once the phases are validated. Since recipe procedure code is decoupled from equipment phase code, the need to revalidate a recipe procedure does not necessarily mean that all phases have to be revalidated. So, theoretically, you can validate new additions to your process or revalidate changes to your process faster. This results in less downtime, faster time-to-market, and all those other great benefits that companies yearn for. (Did you like our repeated use of the root word *validate*? Then you'll love our next book, about automating parking systems.)

We're not going to explain anything more about validation. Do what your company procedures require. (Or if you don't have formal procedures, do what you think is best for your company.)

Start-up Tips

Start-ups can really be a lot of fun. We tried our best to always make them fun at Ben & Jerry's. Unfortunately, start-ups are at the end of a project. If a project is behind schedule for any reason, people are stressed, and the start-up may be rushed. Here are some of the things we learned during the start-up of our batch control system, as well as from the many other system start-ups at Ben & Jerry's:

- *Safety first*—Things will get hectic during testing, and it's only a normal tendency to rush things toward the end of a project. So put people safety at the top of your list, and keep a sharp eye out for issues.
- *Stick with one leader at all times*—The leader will need to serve as a facilitator to discuss countermeasures for problems or to make a retest or continue decision. Make sure everyone agrees on who the leader is. Find an alternate if the leader cannot always be present.
- *Keep the testing team focused and make sure everyone knows his or her responsibilities*—Confirm the testing schedule with your team, especially if testing will occur over a weekend. Nothing is worse than team members not showing up, chaos during testing, or people standing around looking for something to do.
- *Hand out test plans to everyone*—This helps keep everyone focused and serves as a schedule during the start-up process.
- *Don't skip steps*—No matter how well things are going during your testing, follow your test plan. (During the initial plant start-up, we just about doused one of our OEM employees with corn syrup because he insisted on cutting testing short to go home early.)
- *Make use of water testing if your manufacturing process allows it*—We're not advocating that you waste water, but water is not corrosive, it doesn't smell, it generally is not expensive, and it doesn't explode if a spark occurs.

- *Put plenty of time in your project schedule for testing*—Things go wrong, even if the system is perfect. People become ill. Planes are late. Cars break down. Power fails. You get the picture. Start as early as you can.
- *Remember that automation start-up testing is at the end of a project*—If your projects are like ours, particular activities within an overall schedule might take more time than estimated, but your customers are very reluctant to move the final delivery date. So, as the automation professional, many times you're stuck taking the heat for a late start-up. The best way to handle this is to stay on top of the project before your particular responsibilities start. If you see an impact on the overall schedule because of problems earlier on, make it your job to ensure that everyone is aware of the risk of a late start-up. (You can always try to pad your part of the schedule when estimating the project work, but you had better be sharp about it. You may eventually be caught, and either your manager will unpad all future schedules for you, or you will lose credibility with your site.)
- *Don't forget to order pizza for everyone during weekend work*—This adds an appreciative, human element to your project, but it also has a sneaky, hidden motive: it keeps you in control of the lunch break.
- *Give a round of applause after everything is done*—An e-mail to team members' bosses is a simple act of recognition. During some projects, we created large posters to recognize team members after we reached a significant milestone.
- *Learn from your experiences*—To really take advantage of S88 modularity so you can eventually reuse equipment phase code and recipe procedures, it's important that you pass on what you learn in your project. Try creating postproject reports to be distributed throughout your site, division, or company that explain successes, failures, what was learned, and what could be done differently.
- *When you're done for the day, don't allow yourself to be the only one with a clean uniform*—Once others see that, it won't be clean for long.

You see, there really is not all that much to validation and starting up your system. Yeah, right. Okay, let's wrap up this book.

14

FINIS

Hey, this is our last chapter. *Sniff*. . . We'd like to take this opportunity to discuss some miscellaneous, yet very important points.

WHAT WE LEARNED—THE BIG PICTURE

Applying S88 really opened our eyes to the true meaning of batch control. Big process companies, such as petrochemical or pharmaceutical firms, have been refining (no pun intended) their batch control expertise over the decades and typically have a good grip on the subject matter. But there are so many smaller companies—sometimes young companies like Ben & Jerry's—that have yet to leverage S88 for competitive advantage.

Right at the start in Chapter 1, we listed many of the business and technical benefits that result from deploying an S88 batch control system. The following list represents some of the major points we learned implementing our S88 solution:

- *The fact that S88 is an international standard is a big selling point*—Following a given standard is all the rage nowadays. Some technology standards are de facto, such as in desktop computers and operating systems. Adhering to a standard generally minimizes the risk that your solution will not be supported in the future.
- *However, don't feel that you need to explain all the details of S88 to everyone*—Management really cares about productivity improvements, cost reductions, and enhanced quality. At many companies, how the automation professional goes about achieving these doesn't matter to them.
- *If you work for a user company, make sure your vendors are aware of S88*—In today's environment, "user" companies are partnering more with their vendors. Increasingly, user companies are outsourcing much of their detailed control system expertise. To wring out the maximum benefit S88 can provide, your vendors need to understand and make use of the standard as much as you do. (If you work for a vendor company, S88 is another great way of providing competitive advantage to your customers.)
- *Don't underestimate the tangible and intangible benefits of making your operators' jobs easier*—Eliminating unnecessary paperwork or automating a

manual process frees operators up for more important and value-added tasks. Our operators hated paperwork and the manual keying of data. When we eliminated some of it, they loved us even more. If you're in a union environment, you may have some specific challenges to overcome when eliminating tasks, but sell hard. Ben & Jerry's never, ever considered firing or laying off someone just because we could eliminate a job or a portion of a job, so our operators were easy to work with on these productivity improvements. For us, improving our operators' (our customers') work lives was what excited us the most. About two weeks after successfully starting our system, we attended a conference. Since we had left the "old" batch control system in place, we told our operators that if the new system failed while we were gone, just go back to using the old system until we return. They paid us the highest possible compliment by replying, "But we don't want to use the old system."

- *If you're retrofitting a system, don't just replace what you already have*—Take this opportunity to find out how to make things better. Understand how your operators currently run the system, and get their input on how they would improve its operation and their interaction with it.
- *If you're retrofitting a system, don't just replace what you already have*—In this case, you may not wish to trash your old system. Keep it around for a while at least. You may need to temporarily revert to it during the ever-so-tricky start-up phase.
- *Compared to what we went through with control systems before S88, validation was a breeze* — Chapter 13 talks all about this.
- *Be prepared to plug the benefits of the S88 system to your management*—Oftentimes it may not be obvious that the control system is responsible for improvements to your site. For example, on one occasion at Ben & Jerry's, OpenBatch reports allowed us to track a discrepancy between a plant master recipe and a financial recipe standard. If it had gone unchecked, the company could have unknowingly absorbed a material variance of hundreds of thousands of dollars over the course of a year. Our plant controller pointed out to management that OpenBatch provided the key information for us to identify and find the root cause of this issue.
- *Understand that an S88 batch control system is just one element in a total system of manufacturing*—Don't just look at this from a technical standpoint; realize the business side of this statement as well. Two of the biggest business challenges that manufacturing companies face today are successfully planning production and conforming to the plan. The fallout from handling these activities poorly includes missed shipment dates and underutilizing plant resources, which results in second-rate customer service metrics and higher manufacturing costs. A prerequisite to good production planning and conforming to the plan is a solid production execution. Irrespective of good production planning software or well-trained employees, inherent variability in your processes can create havoc. S88 techniques and solutions can help reduce the variability in your

batching operations, allowing you to better predict the time and resources needed to execute an order. So beyond all the benefits we listed in Chapter 1, S88 batch control fits well within your plant or company's total suite of business solutions.

People often ask what was the hardest thing about implementing our S88 solution at Ben & Jerry's. Our answer to this is from a technical perspective. It took some getting used to write PLC phase logic to be state-oriented, including having separate code for the *Running*, *Holding*, *Restarting*, *Stopping*, and *Aborting* transient states. Once we made the transition to this new way of writing PLC logic, however, everything went really well. In fact, maintaining phase logic was a lot easier than maintaining traditional code. Writing subsequent "add-on" phases was practically a joy.

On a more personal note to our peers, keep in mind that developing S88 skills is great for career development. Not only will you become more valuable to your company, but those skills are also easily transferable as you move to other departments or into engineering management. If the need arises, you will find that your S88 knowledge also makes you more marketable to other firms inside or outside your core industry.

A Challenge to Think Beyond Manufacturing

Throughout this book we've kind of made one basic assumption: S88 is for batch control of manufacturing processes. But think about it. Why should S88 apply only to manufacturing? So many things in life occur in batches.

How about theme park attractions? Instead of combining ingredients according to some recipe, you're whipping people around according to some sequence. Why couldn't *Alien Encounter* at Disney's Magic Kingdom be under the control of an S88 batch system? Along those same lines, Epcot Center's nightly *Illuminations* fireworks extravaganza is clearly a batch process.

Would power generation be a candidate for S88? Perhaps. Thinking of producing power in batches doesn't seem to make sense, but the state-based procedural control S88 defines could be a great asset to running a power plant.

What about moving baggage at an airport? The obvious batch is the group of luggage in the cargo hold of a plane. We even know of a diamond company that uses an S88 software product to help sort its gemstones.

For More Information

Now that you're at the end of the book, we want you to feel very assured that you won't simply be cut loose to implement S88 batch control on your own. There are many sources of information and support regarding the standard. Interested in learning more? Here's what we recommend:

- First, make use of all the resources ISA has to offer. Make sure you attend the annual ISA Expo or ISA Tech show held each fall. Your local ISA chapter may have regular meetings where you can share information with your peers. ISA also sells other batch control books and offers S88 classes that you may wish to explore. Check out www.isa.org for more information.
- Next, consider joining the World Batch Forum (WBF). It's a nonprofit professional organization established in 1994 to promote the exchange of information related to the management, operation, and automation of batch process manufacturing. The WBF is an association of end users, vendors, consultants, and academics with a noncommercial agenda. In addition to hosting an annual conference with formal presentations and technical papers, the WBF provides organization, management, and structure to facilitate networking among its members as well as the sharing of knowledge and information related to all aspects of batch processing. You can contact the WBF at www.wbf.org.

- Third, your favorite S88 batch software provider will have the skills and experience to assist you with your projects.
- Finally, there are some very qualified independent consultants who possess a wealth of experience with S88. Some have been members of the S88 committee for years. It would be difficult for us to list them all here, but pay attention to who's speaking at events like ISA shows and the World Batch Forum's annual "Meeting of the Minds" conference. You'll quickly figure out who's who in the world of batch control.

One Last Thing

Given our light writing style, you may be thinking that the title of the book is somewhat conservative. Although we like the title, it wasn't as catchy as the other suggestions we submitted to ISA. So we feel it might be fun to share with you the top ten book titles rejected by ISA. Ready . . .

10. *Seven Habits of Highly Successful Batch Control Systems*
9. *Those Magnificent Men and Their Batch Control*
8. *The Soul of a New Batch Control System*
7. *Who Said Batch Control Was Boring?*
6. *Everything You Always Wanted to Know about Batch Control, but Were Afraid to Ask*
5. *The Joy of Batch*
4. *Everything We Know about Batch Control, We Learned in Kindergarten*
3. *What Harvard Doesn't Teach You about Batch Control*
2. *Midnight in the Control Room of Good and Evil*
1. *Who Would Have Thought That Implementing Batch Control Could Be This Much Fun?*

Thanks for reading, and good luck with your S88 implementations.

INDEX

About the Authors

Jim Parshall is a Senior Process Automation Engineer at Eli Lilly and Company. He is currently assigned to the Indianapolis Dry Products site, which serves as a development center for new tablet and capsule products and provides manufacturing capacity for worldwide distribution. After coordinating the development of an automation and information technology strategy for the site, Jim is now planning strategic automation and information integration projects. He leads the corporate team responsible for overseeing a global supplier alliance with Rockwell Automation and is a member of the company's global process automation strategy team.

Before joining Lilly, Jim worked as an automation engineer at Ben & Jerry's Homemade in St. Albans, VT. He led the design and delivery of process control systems during the exciting startup of the company's newest and largest plant. As part of his process control and supervisory system responsibilities, Jim led the successful installation of an S88-aware batch control and management system.

He received his BSEE from Kettering University (formerly General Motors Institute), and received his MSEE with emphasis in manufacturing from Purdue University. Jim is an active participant in the World Batch Forum, and he is a member of the ISA SP95 standard committee on enterprise/control integration.

Jim, his wife, Georgianne, and their son, Evan, live in the Indianapolis area.

Larry Lamb is the Software Manager at Oakes Electric, a New England/New York Rockwell Automation distributor. He is responsible for leading the business and technical aspects of providing technology-based solutions to customers.

Before joining Oakes, Larry was a controls engineer at Ben & Jerry's Homemade in St. Albans, VT. His expertise drove the design of the control architecture at the new plant in St. Albans, and Larry was responsible for writing the process control and operator interface code for the plant production area.

Larry began his career as a licensed pharmacist working in hospitals and retail for 11 years. His interest in computers led him to become an electrical engineer. Larry received his BS Pharmacy from the Massachusetts College of Pharmacy, and received his BSEE from the University of Massachusetts in Amherst. On occasion, he continues to work in retail pharmacies to maintain his license. Even though he no longer works for Ben & Jerry's, he enjoys being mistaken for Ben.

Larry and his wife, Deb, have two daughters, Erica and Sarah. They live in northern Vermont.